Technology and Human Capital in Historical Perspective

Also by Jonas Ljungberg

THE PRICE OF THE EURO

Technology and Human Capital in Historical Perspective

Edited by

Jonas Ljungberg
Lund University
Sweden

and

Jan-Pieter Smits
University of Groningen
The Netherlands

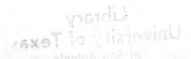

First published 2004 by
PALGRAVE MACMILLAN
Houndmills, Basingstoke, Hampshire RG21 6XS and
175 Fifth Avenue, New York, N.Y. 10010
Companies and representatives throughout the world

PALGRAVE MACMILLAN is the global academic imprint of the Palgrave
Macmillan division of St. Martin's Press, LLC and of Palgrave Macmillan Ltd.
Macmillan® is a registered trademark in the United States, United Kingdom
and other countries. Palgrave is a registered trademark in the European
Union and other countries.

ISBN 1–4039–2067–2

This book is printed on paper suitable for recycling and made from fully
managed and sustained forest sources.

A catalogue record for this book is available from the British Library.

Library of Congress Cataloging-in-Publication Data
Technology and human capital in historical perspective/edited by Jonas
    Ljungberg and Jan-Pieter Smits
        p. cm.
    Includes bibliographical references and index.
    ISBN 1–4039–2067–2 (cloth)
    1. Technological innovations—Economic aspects—History.   2. Human
    capital—History.   3. Economic history.   I. Ljungberg, Jonas.   II. Smits.
    Jan-Pieter
    HC79.T4T413 2005
    338'.064—dc22   2004049121

10   9   8   7   6   5   4   3   2   1
13   12   11   10   09   08   07   06   05   04

Printed and bound in Great Britain by
Antony Rowe Ltd, Chippenham and Eastbourne

Contents

List of Figures

List of Tables

Preface

Many colleagues have contributed to the present book. It is a produce of a network around issues related to long-term economic growth. Some networkers, among them the two editors of this volume, discussed the topic of technological change and economic growth at the conference 'Technology and productivity in historical perspective', which was organized in Wassenaar (The Netherlands) in 1999 under the auspices of the Posthumus Institute. Over time this network has substantially broadened, and had one stepping stone at the XIII International Economic History Congress in Buenos Aires, summer 2002, where we organized a session from which the book has inherited its title. Traditionally such sessions used to be prepared through a pre-conference but in this age of new information technology an e-dialogue, including comments by anonymous referees, filled that purpose. Most chapters have an origin in the Buenos Aires session, although a few have materialized afterwards. In preparation of the book the authors and editors gathered at a workshop outside Lund, in the autumn 2003, financed by the Bank of Sweden Tercentenary Foundation, which also has supported the Lund research project 'Economic Growth and Productivity in a European Perspective since 1870'. This project, headed by Lennart Schön, motivated, and made out much of the basis, for the networking. Now, when this book comes off the print, we would like to acknowledge our gratitude to those networkers who in various ways have contributed in the process: Bart van Ark, Steve Broadberry, Erik Buyst, Herman de Jong, Jan Tore Klovland, Pedro Lains, Anders Nilsson, Albrecht Ritschl, Max Schulze, Lennart Schön, Ken Sokoloff, Hans Voth. Finally, thanks to Kyla Madden, Kingston, for her daunting effort to brush up the style and to Astrid Lieng, Lund, for standardization of the format of the diverse chapters.

Lund University   JONAS LJUNGBERG

University of Groningen   JAN-PIETER SMITS

Notes on the Contributors

Maria Mar Cebrián is a PhD student in the Department of History and Civilization, European University Institute, Florence, Italy.

Camilla Josephson is a PhD student in the Department of Economic History, Lund University, Sweden.

Svante Lingärde is a PhD student in the Department of Economic History, Lund University, Sweden.

Jonas Ljungberg is Associate Professor and Director of Studies in the Department of Economic History, Lund University, Sweden.

Santiago López is Associate Professor in the Department of Economics and Economic History, Salamanca University, Spain.

Marvin McInnis is Professor Emeritus of Economics, Queen's University, Canada.

José Ortiz-Villajos is Associate Professor in the Department of Economic History, Complutense University of Madrid, Spain.

Jaime Reis is Senior Research Fellow, Instituto de Ciências Sociais, University of Lisbon, Portugal.

Jani Saarinen is a PhD student in the Department of Economic History, Lund University, Sweden.

Jan-Pieter Smits is Associate Professor in the Faculty of Economics, University of Groningen, The Netherlands.

Lars Svensson is Associate Professor in the Department of Economic History, Lund University, Sweden.

List of Abbreviations

CIS	Community Innovation Survey
CPI	consumer price index
CR	cointegration relation
CSIC	*Consejo Superior de Investigaciones Científicas*
DCB	*Dictionary of Canadian Biography*
EMS	European Monetary System
FNICER	*Fundación Nacional para Investigaciones Científicas y Ensayos de Reformas*
GDP	gross domestic product
GNP	gross national product
HTE	higher technical education
ICT	information and communications technology
IPC	International Patent Classification
ISIC	International Standard Industrial Classification
IVIE	Instituto Valenciano de Investigaciones Económicas
JAE	*Junta para la Ampliación de Estudios*
LM	Lagrange Multiplier
LO	Confederation of Labour (Sweden)
OLS	ordinary least squares
R&D	research and development
RCA	Revealed Comparative Advantage
RMC	Royal Military College (Canada)
RPA	Revealed Production Advantage
RRDA	Revealed R&D Advantage
RTA	revealed technological advantage
SAF	Swedish Employers' Confederation
SIC	Standard Industrial Classification
TAF	Technology Assessment and Forecast (by USPTO)
TFP	total factor productivity
USPOC	United States Patent Office Classification
USPTO	United States Patent and Trademark Office
VAR	vector auto-regressive
VF	Engineering Employers' Association (Sweden)

1
Technology and Human Capital in Historical Perspective: An Introduction[1]

Jonas Ljungberg

Few would deny that technological change has transformed both production and everyday life since pre-industrial times. The implications of this statement produce divergent opinions, however. Even if it is mostly agreed that everyday life has changed for the better, opinions differ on how technology has altered production, and continues to do so. For example, the inter-play between technological change and economic growth has been labelled a 'black box' (Rosenberg 1982; 1994). It is difficult to know what goes on inside the black box, and it is hard to find factors that determine technological change, or at least to find a linear relation between certain factors and the economic performance. Throughout history efforts have been made to incorporate technological or technical change into economics, but no coherent theory has earned universal consent.[2] Mainstream economics have instead regarded technology as a given exogenous factor that greatly influences the economy but which is not itself affected by the economy. Recently this has changed among economists with the emergence of endogenous growth theory. For economic historians, however, technological change is an old issue. The present book is an attempt to combine different approaches to technological change. Common to these approaches is the dimension of time, whether nineteenth-century history or the more recent past. The reference to 'historical perspective' in the title is an allusion to Alexander Gerschenkron (1952) and his impact on economic history.

In the preceding paragraph we can begin to see some oscillation between 'technical' and 'technological'. Technology refers to a system, theory or paradigm and a deeper knowledge, whereas technique refers to a practice or practical knowledge. The distinction is not easy to make, however, and in the literature 'technology' has come to embrace both meanings.

A reference to 'human capital' is also made in the title. Occasionally human capital is subsumed under technology, as it is the most important factor behind technological change. However, their non-linear relation, as well as the many aspects that are connected with the understanding of human capital, necessitates its emphasis. A controversial question,

1

approached in several of the chapters, concerns the implications of technical change for labour.

It is sometimes remarked that Marx, contrary to older classical economists, recognized technology as a key factor in economic development with his concept of 'productive forces'. Nevertheless, the actual innovator or agent of technology was largely ignored by Marx who saw the impersonal accumulation of capital as the driving force. The conditions that fostered technological change escaped his notice, as he was more concerned with the effects of the new technology on labour. The artisan controlled his tools but in factory production the worker was turned into an accessory of the machine. This view has been developed by Braverman (1974) who saw the twentieth century as a period of continuous de-skilling and degradation of both blue-collar and white-collar labour.

Although economists have long agreed on the importance of education, the theory of de-skilling has not been thoroughly scrutinized. On the contrary, when physical capital was substituted for labour with mass production, the de-skilling theory was not questioned. About 1980, however, when earnings differentials between professionals and workers, and thus the premium on education, broke a long-lasting downward trend, the issue was raised from a new angle. Increasing earnings differentials are an indication of a higher demand for knowledgeable and skilled labour. It was concluded that the new technology, developing on the basis of the microprocessor, complements skilled and professional labour (see, for example, Katz and Murphy 1992; Doms, Dunne and Troske 1997).

In an important article, Goldin and Katz (1998) traced the complementarity between technology and skills to the early twentieth century. According to their argument, technology and skills were substitutes during the Industrial Revolution but turned into complements after the introduction of electricity during the Second Industrial Revolution 'because it reduced the demand for unskilled manual workers in many hauling, conveying, and assembly tasks' (p. 695). The main reason for the shift from substitutability to complementarity is, however (according to Goldin and Katz), explained by the difference between production workers on the one hand, and installation and maintenance workers on the other. The latter require a specific set of skills, whereas the former might be less skilled or unskilled.

To delve further into this issue, it is necessary to redefine the problem. First, it is difficult to escape the fact that technological development is based on the progress of knowledge. Second, the greater the division of labour, the more diversified the demand for skills will be. Knowledge is a broad concept that comprises science and technology as well as skills, the latter denoting hands-on or practical capability. Rosenberg (1994, p. 28) has drawn attention to the explanation of the role of skills by Charles Babbage, who was Marx's precursor on the question of technology, in a treatise published in 1833. Babbage's point was that the division of labour did not simply mean that the production process was chopped up into a series of trivial moments, resulting in greater output. The economy of the division of labour is that the diverse moments of the production process are performed by workers with no more skills than necessary. In this system, the skilled worker or the engineer is only paid for work that requires certain skills, and is not assigned unskilled work (1833, pp. 175–6).[3]

Babbage also raised the issue of the effects of machinery on employment, and not only as far as unskilled or raw labour was concerned. His discussion remains a relevant research plan as regards substitutability versus complementarity between technology and skills:

> the circumstance of our not possessing the data necessary for the full examination of so important a subject, supplies an additional reason for impressing, upon the minds of all who are interested in such inquiries, the importance of procuring accurate registries, at various times, of the number of persons employed in particular branches of manufacture, of the number of machines used by them, and of the wages they receive.
>
> (Babbage 1833, p. 337)

Since then, the growing share of professionals in the labour force suggests that de-skilling is not a general characteristic of industrial production. The Swedish development is typical for industrialized countries: in the 1890s, white-collar labour amounted to only 4 per cent of the labour force, but by the 1960s it had expanded to more than one-quarter of the labour force in manufacturing (Ljungberg 2004a). If the division of labour has involved white-collar labour its expansion can be seen as a measure of the growing importance of knowledge. Knowledge is a much wider concept than technology, but technology is a key part of a society's knowledge base.

The Knowledge Society has become a catchword that should have emerged recently with the new information technology. However, it could be disputed whether this is a difference in quantity rather than in quality. In *When Information Came of Age*, Headrick (2000) argues for the decisive importance of the advance of knowledge in early industrialization. It was not a coincidence that the Industrial Revolution occurred alongside the breakthroughs of the 'Age of Reason'. Mokyr (2000; 2002) has further analysed the role of knowledge in economic growth, and his conceptual framework also addresses the role of technology.

Many efforts to capture knowledge, technology, innovations and so forth in a theoretical framework have resulted in concepts or metaphors that are of limited analytical value. Here Mokyr (2000; 2002) supplies a helpful tool box. Starting from Kuznets's (1966, pp. 86–7) emphasis on the role of 'useful knowledge' for modern economic growth, Mokyr draws a line between prescriptive knowledge and propositional knowledge. The former is about how to do things, such as techniques and rules of thumb. Propositional knowledge explains why things work in the way they do. A discovery is thus an addition to propositional knowledge, while an invention adds to prescriptive knowledge. This categorization of useful knowledge is more fruitful than the conventional distinction between 'scientific knowledge' and 'empirical knowledge', which is unclear about the functional difference.

In an institutionalist perspective the importance of intellectual property rights, such as patent legislation, is often stressed. Of no less importance, however, was the emergence of open science from the seventeenth century, transforming science from more or less proprietary secrecies to a public good. The importance of this change becomes clear when we look at China, where science continued to be the preserve of the mandarins. In Europe, property rights and secrecy applied to prescriptive knowledge, whereas propositional knowledge entered the domain of free exchange. The exception to this partition is that prescriptive knowledge to some extent also entered the public domain, and was diffused by enlightened societies and printed publications. Scientific method, scientific mentality and scientific culture characterized what Mokyr calls 'Industrial Enlightenment' (2002, pp. 35–6).[4] There is plenty of historical evidence for how an institutional framework for the communication and diffusion of knowledge grew in the era of Enlightenment. According to Mokyr: 'The interaction between propositional and prescriptive knowledge grew stronger in the nineteenth century. It created a positive feedback mechanism that had never existed before' (2002, p. 117). Institutions of the *ancien régime*, like guilds, had in the preceding centuries posed a negative feedback that aborted the accumulation of useful knowledge.

Prescriptive knowledge, even if it is successfully applied, must not be based on 'correct' propositional knowledge. However, it is difficult to imagine that the production of iron and steel, chemical stuffs, steam and water power or medical services could have developed very far without a scientific method, a scientific mentality and a scientific culture, and finally, without the further development of science. Rosenberg has dealt with the feedback from technology to science, highlighting, for example, how the aircraft industry has pushed the science of turbo dynamics further ahead, or how the early railways pushed the science of metallurgy (condensed in 1982, ch. 7). Of course the feedback travels in both directions, and the more it develops the more endogenous technology becomes in the economy. Positive feedback fosters not only the growth of useful knowledge but also breaks the vicious circle between population growth and backlash which had trapped pre-industrial growth.

Here an obvious connection arises with endogenous theories of economic growth. Without an accumulation of propositional knowledge, prescriptive knowledge will eventually generate diminishing returns, simply because every producer will have adopted the same technology and no one will be able to push it further ahead. However, as with all hypotheses of grand theory, the question is how to test it. Which steps should research take according to the new framework? Which unresolved problems could it shed new light upon?

The main avenue for the analysis of economic growth has developed along the production function which, in different versions, derives change in output from the change in input of production factors. Ever since Abramovitz (1956) highlighted the size of the residual of the production function, and with a now famous phrase called it 'some sort of measure of our ignorance', economists have searched for elements that can explain the changes in productivity of capital and labour. There are the proximate factors contributing to total factor productivity (TFP) which are analysed in growth accounting, but there are also the ultimate factors, such as institutions and international economic regimes that are more often featured in economic history and development economics.

Technological change was the first explanation for TFP (also known as the residual). Later, Denison (1962) pioneered the growth accounting attempt – taken up by, for example, Maddison (1987) – to decompose TFP into different elements such as economies of scale, foreign trade effects, education, structural change, windfall effects of new natural resources, and so forth. This is one way to qualify the content of technological change and to reduce our ignorance through analysis of the residual. However, another line of thought perceives the residual as an error of measurement. Is technological change disembodied or embodied in capital? If disembodied, then technological change is reflected in the residual; but if quality change of capital can be embodied in the quantity data of capital, then the residual might be drastically reduced. Jorgenson (1966, 2001) and Jorgenson and Griliches (1967) therefore try to qualify capital in taking account of vintage effects, and substitute capital services for the traditional capital stock data. In a similar way, the increased level of human capital in the labour force is included in the production function. As a result, TFP almost disappears and economic growth is seen as strongly driven by capital goods, and to a lesser extent by human capital. Another approach, endogenous growth theory, still assumes technological change as disembodied in physical capital, but includes human capital in the production function, and as a result produces a smaller residual. Knowledge is here understood as the fundamental determinant of technological change, and proxied by, for example, schooling or learning-by-doing (Lucas 1993) or research and development (R&D) investments (Romer 1990).

In their search for the sources of growth, economists and economic historians have thus attempted either to decompose or to squeeze TFP, or the residual, as much as possible. It remains to be seen if technological change can, or should, be fully endogenized. In particular, major innovations such as the steam engine, the harnessing of electricity or the microprocessor are impossible to predict and therefore difficult to treat as endogenous in a model. As Mokyr has observed, endogenous models 'have attempted to open these black boxes, but have just found another black box inside' (2002, p. 116). One might think that learning effects (learning-by-doing, learning-by-using), which are crucial in bringing about increases in productivity, would be more apt to endogenize. However, as shown in this book by Camilla Josephson in Chapter 7, such assumptions should be treated as hypotheses and tested before conclusions are drawn.

Black boxes notwithstanding, the unpredictable role of genius in technological change misleads Mokyr into the assement that 'it is not necessary for many people to have access to the epistemic base' (2002, p. 14): that is, to propositional knowledge. The prerequisite of only a limited distribution of knowledge is inferred *ex post* in an account of the British Industrial Revolution during which 'a small group of at most a few thousand people ... formed a creative community based on the exchange of knowledge' (2002, p. 66). *Ex ante*, however, the broader the distribution of the propositional knowledge and thus the level of human capital in any given society, the greater the probability that path-breaking discoveries and inventions will emerge.

Literacy is one standard measure of human capital, and its historical role illuminates the point. The established view is that the achievement of literacy was of little importance to Britain's Industrial Revolution. Evidence in support of that view includes the insignificant change in literacy rates in the half century before 1830. It is also supported by the census of 1841, from which it has been concluded that less than 5 per cent of workers were in occupations where literacy was necessary (Mitch 1993). Certainly this is an indication of the low level of human capital, as far as formal education is concerned, that was demanded for factory workers in the early stages of British industrialization. However, it tells us little about the role of literacy among other actors in the economy, and nothing about the general importance of literacy. It might be helpful to distinguish between human capital and social capital. Human capital belongs to the individual, and should take account, for example, of the use literacy has for the individual, while social capital considers the use of literacy for the society at large. Social capital is an institutional factor that influences the way society works; it is one of the components of social capability, to use Abramovitz's broad concept (1979; 1995). As regards actors other than factory workers, research on early nineteenth-century Swedish agriculture has shown that functionally literate peasants were more successful as entrepreneurs. Literacy is found to have been an efficient transaction technology during that period of transformation and commercialization of the economy (Nilsson, Pettersson and Svensson 1999; Svensson 2001). As regards social capital, more research is needed on the impact of the expansion of primary education that preceded a successful performance in the Second Industrial Revolution in some countries. A reasonable hypothesis is that the more enlightened the population, the more smoothly will a country adopt innovations and structural transformations. There is also a greater probability that path-breaking discoveries and inventions will occur.

Recent medical research on the role of literacy for the development of social abilities points in the same direction. An amazingly well-adapted population for such a study was found among fishing families in Portugal. The eldest daughter traditionally was not allowed to go to school but stayed home and helped her parents with household production. The researchers could thus compare a group of illiterate daughters with their literate siblings, who otherwise had experienced similar conditions. They found that the ability for systematic thinking and memorizing is significantly improved with literacy (Petersson, Reis and Ingvar 2001).

A crucial point for the importance of useful knowledge, as suggested by Mokyr, is the interpretation of the British Industrial Revolution. Revisionists have criticized the Rostowian view of industrialization as a sudden 'take off' and argued that it was a more gradual process. According to the Crafts–Harley view, the change that accompanied the rise in productivity was limited to the 'modernized sector' (textiles, iron and transport), which represented one-fifth of the economy. Agriculture, representing another fifth, also contributed to productivity change, although less so. The remainder, denominated the 'traditional sector', did not contribute at all (Crafts and Harley 1992; Harley 1993; Harley and Crafts 2000). Even if economic growth is highly dependent on leading sectors that pull the transformation of the whole economy ahead, one might doubt the Crafts–Harley characterization of the sectors during the early stages of British industrialization stretching out over half a century or more. It is difficult to square the concept of modern economic growth, distinguished by its sustainability from previous bursts of growth in history, with a technical change that for so long a period was restricted to certain branches of manufacturing and transport. Where is the role for the feedback mechanism (information) between propositional and prescriptive knowledge,[5] if change was so isolated?

At the crossroads of modernity and technical change in production was the printing trade. Important innovations were introduced in the early nineteenth century and with a daily newspaper, *The Times*, as partner and pioneer. In 1800, iron substituted wood in the printing press, and just over a decade later printing could be powered by steam instead of muscular energy. The number of copies printed in an hour doubled, and by 1828 had increased another four times, to 4,000. Publishing enterprises became profitable and mushroomed. 'Between 1820 and 1840 at least 2000 new newspapers and periodicals, many with illustrations, made their appearance in Britain, and in other countries there was similar enterprise' (Berry 1958, p. 702). Lithography also emerged in this period, leading to an increase in illustrated matter, from cartoons to constructions. Babbage was a contemporary in this development, and gives a vivid account, with a clear consideration of the distribution of useful knowledge (1833, pp. 269 ff.). Even if Jeremy Black (1994, p. 39; 2001, p. 72) plays down the transformation of the British press during the Industrial Revolution, in comparison with the changes before

1750 and after 1850, the evidence of quantitative expansion is striking, particularly considering the number of newspapers and volume of paper consumption. Taxes on paper and advertising was a constraint for the expansion and probably resulted in a small scale pattern instead of fewer big dailies. Continuous paper and rotary printing were available technologies from the early nineteenth century, but the consumption taxes meant high prices and acted as a brake on mass production. By mid-century these taxes were lifted and in the 1860s printing presses were fed by newsprint on a roll (Berry 1958, p. 699; Black 2001, pp. 178–9).

Given the growing streams of information, themselves a result of technological change, one might wonder if that development could remain isolated in a few sectors of the economy. Improvements in transport are commonplace in narratives of the Industrial Revolution, and it is also reckoned as part of the modernized sector. According to the estimates by Feinstein (1988, p. 444), the growth rate of investments in transports can be calculated to 1.5 per cent annually between the 1760s and 1820s, which certainly yielded productivity gains. However, the most rapid growth of investments, 2.6 per cent annually in the same period, was in the construction of dwellings. It could be that this was a case of extensive growth without productivity change, but this seems implausible. It is deceptive to think of 'traditional' trades as leaving no room for productivity change. For example, Pollard (1981, p. 79) noted that labour productivity in British brick production during the Industrial Revolution was half that attained by brickworks in the Netherlands in the seventeenth century. Did British brick-making remain that far behind? It is not unreasonable to presume that the potential for catch-up in that 'traditional' area of production was realized when information about production methods could flow more freely. In the four years between 1821 and 1825, British brick production more than doubled (Angus-Butterworth 1958, p. 373). Was that a result of greater inputs with no residual of improvements in practice, organization and distribution? Glass is an associated buildings material also belonging to the 'traditional' sector. Since the technology for the production of flat glass was pioneered in France and Germany, how could Britain manage to achieve a competitive advantage in the manufacturing of glass plates around 1830? That such a development took place can be inferred from the considerable price differential between large glass plates that Babbage (1833, p. 159) quoted in London, Paris and Berlin.

Even without a stagnant 'traditional sector', productivity change and economic growth during the Industrial Revolution might well have been much more gradual than the first quantitative estimates indicated. The experience of the Solow Productivity Paradox in both the Second and Third Industrial Revolutions teaches us that nothing else should be expected.[6] The gradualist interpretation, most distinct in the Crafts–Harley view, evolves around the question of the origins and driving forces of industrialization. Without diminishing the importance of quantification and growth accounts, a new focus of research emerges with the emphasis on the role of knowledge. 'The true question of the Industrial Revolution is not why it took place at all but why it was sustained beyond, say, 1820' (Mokyr 2002, p. 31). Is the answer, as Mokyr suggests, that it was the advance of propositional knowledge and the new institutions for the feedback mechanism with prescriptive knowledge which proved to be the decisive difference compared with previous historical periods? If so, then the importance of historical research in the fields of technological change and human capital for the understanding of modern economic growth has once more been demonstrated.

* * *

The studies collected in this volume have three different themes. The first concerns the role of human capital in early industrialization up to the present. The second analyses the residual (that is, TFP) during the last half century. The third concerns patents used as a proxy for technological performance, an approach applied for different historical periods and countries. Each of the studies concerns a country on the periphery of the North Atlantic economy: Portugal, Spain, Sweden, Finland and Canada are all addressed. That circumstance connects this volume with another recent debate about globalization. Whereas periods of globalization, such as 1850–1914 and the decades since 1945, have experienced economic growth and convergence between nations, periods of de-globalization, such as 1815–50 and 1914–45, have been less prosperous. It is argued that open economy forces, international trade and factor movements are the principal causes of convergence, which does not primarily consider gross domestic product (GDP) per capita but real wages of unskilled labour (O'Rourke and Williamson 1995a; 1995b; 1999; Williamson 2000). Although convergence and economic growth are not identical processes, they have been mixed up. Open economy forces have been offered as a rival explanation for economic performance, instead of internal factors such as technological change or human capital (for a discussion, see Ljungberg 1996; 1997). A reasonable argument is that there is interaction between external and internal factors, and that social capability can explain why some countries catch up and others do not during periods of globalization. The chapters on Spain, which was slow to industrialize, and on Canada and Sweden, both of which experienced swift industrial breakthroughs in the late nineteenth and early twentieth centuries, are cases in point.

Human capital and industrialization

Studies of human capital during early industrialization are still few and far between. Joel Mokyr suggests that the rise of factories 'is inseparable from the growth in the knowledge-base of production' and that the move from cottage or home production to factories was necessary since 'efficient production required more knowledge than a single household could possess' (2002, pp. 131, 139). However, Mokyr advances no further than Babbage on the question of whether industrialization 'on balance raised or reduced the demand for skills' (2002, p. 142). Even if new competence was demanded and the variance of the skill distribution increased, there is yet no final account of the demand for skilled and unskilled labour.

Chapter 2 by Jaime Reis is a contribution to such an account. On the basis of data from an industrial census in 1890, he investigates the level of human capital among manufactory workers in Lisbon. The focus of the analysis is the skill premium, the remuneration above a certain level of pay assessed as 'raw labour'. The wide, and negatively skewed, distribution of wages among the Lisbon manufactory workers is striking. In other words, unskilled 'raw labour' represented a minority, while most occupations in this modern sector of the Portuguese economy required different grades of skill. Human capital and technology were already complementary in this case of industrialization, which occurred fairly late but still before the Second Industrial Revolution. A similar pattern can be observed among workers in developing countries with foreign enterprises: both skills and wages are raised above those of the traditional economy (Lindert and Williamson 2001). However, the evidence offered by Jaime Reis shows that in the early industrialization of Portugal, the new technology increased the demand for knowledge and skills. It is a challenging task to find out whether the Portuguese case is a model or an exception in history.

Engineering expertise and the Second Industrial Revolution

One of the riddles of the 'black box' is the low correlation between the number of engineers in a country and its ranking in economic performance. A century ago, several countries in Europe had proportionately more engineers than the USA or Canada, but North America experienced greater economic growth. Fox and Guagnini (1993, p. 6) contrast the over-supply of engineers in Italy with the higher demand in Germany and the USA, and see industrial development as the independent factor, whereas Donovan (1993) emphasizes the qualitative difference between a conservative European and a professional American technical education. In the discussions at the Klinta workshop that preceded this volume, Marvin McInnis, author of Chapter 3, suggested the concept of 'knowledgeable entrepreneurship' to deal with this problem. The training and the field of technology must be relevant and up-to-date. Moreover, the institutional conditions must allow entrepreneurship. Well into the Second Industrial Revolution many engineers received their training through practice, in a case of learning-on-the-job. Engineering was traditionally a branch of military and infrastructure tasks and was not geared towards industrial enterprise and innovation. 'Knowledgeable entrepreneurship' is somewhat more specific than 'social capability' and could be a tool for understanding the same type of problem: processes in which lin-ear relationships are difficult to establish. The reader may find that concept useful when comparing the Canadian case with, for example, that of Spain.

Marvin McInnis tells two important stories in Chapter 3. The first is about Canadian economic growth in the decades before 1914. The traditional explanation for the extraordinarily rapid growth of the Canadian economy in that period is the wheat boom of the early twentieth century. However, a close inspection of the time pattern as well as the disaggregated data has led McInnis to the conclusion that Canadian growth was primarily generated by manufacturing. Here the second story begins: why was Canada so successful in the Second Industrial Revolution? Without the rapidly expanding profes-sion of engineers, which is portrayed here with many individuals also acting as entrepreneurs, the performance would not have been so noteworthy. The crucial point is that many of them were typical entrepreneurs, venturing into new enterprises and new technology.

Technology shifts and labour market institutions

A problem for Schumpeterian long wave theories is to explain why innov-ations should cluster according to a wave-like time pattern. However, research carried out at Lund, originally for the construction of historical national accounts, has developed a different approach to long waves. It is not a theory but an empirically uncovered pattern in Swedish economic development, where two distinct phases, roughly 20 years in duration, have alternated with each other over the course of two centuries. The pattern can be seen in series on investments and the wage share, in the export ratio, in relative prices, and in several related variables such as credit market policy and relative earnings. The first phase is named 'transformation', character-ized by an increasing investment ratio and a decisive fall in relative prices of goods with a high technological content. The second phase is named 'rationalization', when the investment ratio decreases but the wage share increases, relative prices of the aforementioned goods only fall moderately, but the export share increases. In more qualitative terms, the transforma-tion phase is characterized by changes in both enterprises and institutions, whereas the rationalization phase is characterized by stability and efficiency. Bubbles and financial crises have halted most of the transformation phases, while another type of crisis, demanding both structural and technological changes, has marked the end of the rationalization phases. Both the 1930s and the 1970s witnessed examples of the latter crisis (Schön 1989; 1991; 2000; Ljungberg 1990; 1991). The time pattern is explained by the diffusion of innovations: not purely technological factors, but by the way they are absorbed in the economy.

Dahmén (1970 [1950]) introduced 'development bloc' to describe how new innovations could give rise to complementarity between different actors or sectors. A development bloc often has a typical life cycle in which maturity and ageing claim a new transformation or a close down. Dahmén's development bloc can be twinned with Bresnahan and Trajtenberg's (1995; see further Helpman 1998) General Purpose Technology (GPT): major innovations such as the electrical continuous current system have complementarities that pervade social and economic life, and do so for an extended period of time. For various reasons, roughly twenty-plus-twenty years have demarcated the life cycle of important development blocs or GPTs.[7]

In Chapter 4 Lars Svensson applies the cyclical approach to the technology–skill complementarity problem, as well as to institutional change in the Swedish labour market. While Jaime Reis traces the origin of the technology–skill complementarity further back in time, beyond the Second Industrial Revolution, Svensson argues that the demand for skilled labour has shown distinct variations during the twentieth century. In Sweden, these variations have followed the cyclical phases of transformation and rationalization, because, as explained by Babbage, the skilled worker should not be paid to do the less skilled work that comes out of a process of rationalization. Thus it is another qualification of the Goldin–Katz thesis. Labour market institutions, not least 'the Swedish Model', are often described as the causal factor of the earnings distribution. However, Svensson considers labour market institutions as an endogenous factor, itself largely explained by economic conditions. This chapter elaborates the idea of 'endogenous institutions' that Knick Harley (1991) has extracted from the methodology of Gerschenkron.

Higher technical education

The measurement of human capital, and how to make it operational in economic analysis, particularly in a historical context, is a fundamental problem. For example, a long-term accumulation of knowledge pertains to both the average worker in manufacturing, whose productivity increased about 20 times during the twentieth century, to the engineer who steers production or design innovations, and to the surgeon who performs a major operation without large incisions in the body of the patient. However, if only the years of schooling, or the differential in earnings above that of the average worker, are said to signify the amount of human capital, countries will end up with diminishing human capital per capita over time.[8]

In Chapter 5, Jonas Ljungberg approaches this and other problems of human capital through an exploration of data on higher technical education in Sweden since 1867. The question is whether higher technical education did pay both from a micro and from a macro point of view. To determine this, the expenditures of the technical institutes, the differences in lifetime income between graduate engineers and college engineers, and output estimates of higher technical education are explored. As Charles Babbage remarked, 'the errors which arise from the absence of facts are far more numerous and more durable than those which result from unsound reasoning respecting true data' (1833, p. 156). Even if 'unsound reasoning' should always be avoided, it is clear that despite the massive accumulation of statistical data, there are still some fields in which empirical facts are surprisingly lacking. A case in point is the economics of education and related problems.

Economic growth in Spain

The economic history of Spain is a record of centuries of economic backwardness that finally turned around beginning in 1960. There are various explanations for this, including the lack of human capital, which Landes (1998) traces back to the Reconquista. Tortella (1994) places much emphasis on climate, which slowed technological change in agriculture, and consequently slowed urbanization and the shift of labour out of agriculture. Others claim that institutional conditions, stretching back to the seventeenth century, had similar consequences for agriculture. All the same, the low levels of technology and human capital were a constraint on Spain when much of western Europe underwent industrialization in the nineteenth century. That Spain lagged behind is clear from a comparison with other 'peripheral' countries during the Second Industrial Revolution of 1890–1914: the Spanish GDP per capita grew with modest 0.9 per cent annually (Prados de la Escosura 2003), compared with 2.3 for Italy, 2.5 for Sweden and 3.0 for Canada (Maddison 1995; for Sweden, see SHNA).

The economic failures in Spanish history are the common point of departure for both Maria Cebrián and Santiago López, and José Ortiz-Villajos. Ortiz-Villajos analyses patent data and this is discussed below. Cebrián and López address TFP, which is the theme of Chapters 6 and 7. How could Spain's switch to rapid economic growth and ability to catch up with the richer industrial countries after 1960 be explained? The authors combine their own growth account for Spain during the short but important period of 1960–73 with accounts for other countries available in the literature. Despite the chosen 'embodied capital' methodology, which should take account of quality change in the capital goods that make up investments, the contribution of TFP in Spain was surprisingly important. The suggested answer lies in an intricate combination of growth accounting and examination of the industrial history of the period. Machinery was imported at international prices, but due to the low supply of human capital, Spanish importers also had to purchase foreign technical assistance. Cebrián and López regard this as an increase in capital volume, but could not separate this cost when deflating the series into constant prices. Therefore the contribution of capital is biased downwards, and TFP is biased upwards. With this in mind, the main point about the importance of technology import to Spanish post-war catch-up is convincingly established.

Variations in TFP

Those who are interested in the analysis of economic time series have experienced the frustration that most economic data display a trend, or in the now established language, have a unit root that defeats conventional methods. Broadly speaking, two series are cointegrated if there is a stable linear relationship between them, and can then be analysed as if there were no trend. However, the lesser the transparency in any method, the greater the care with which it must be handled.[9]

Chapter 7, where Camilla Josephson applies a cointegrated vector autoregressive (VAR) model, is innovative in at least two respects. One is the way each step is made as transparent as possible to the reader. The conventional econometric enterprise is to define a model with dependent and independent variables, and test their explanatory power and statistical significance. In the cointegrated VAR model the dependence relations are not decided on beforehand, which is why it is particularly useful in historical analysis. Josephson uses this model to test eight hypotheses about the determinants of TFP. Also innovative is the use of growth accounting in a disaggregated analysis of the economy. The conventional growth account deals with the performance of the aggregate economy and not with sectoral developments. Josephson searches, however, for the determinants of TFP in three different sectors of manufacturing according to their use of capital, labour and knowledge. The result also shows that TFP is differently composed in the three sectors. Contrary to endogenous growth theory, one of the findings indicates diminishing returns for human capital in the long term. The time aspect plays an important role in the analysis, and it is demonstrated that during the diffusion of knowledge about a certain technique, exclusive knowledge and path-dependent solutions violate perfect competition. The implications of this are fundamental, both for the economy and for the theoretical model.

Spain's low technological level

One possible method of exploring the inside of the black box is to study series of patent data. As with all measures that attempt to quantify technology, patent data are ambiguous. This is due in part to differing degrees of importance among various patents, and also to differences in legislation between countries and over time. At the sectoral level there are also differences in the tendency to patent or to rely on secrecy. However, one of the points in quantification is that the anomalies that are apparent at the individual level are lost in the average at the level of mass data.

Under-pinned by a meticulously collected database of patents, José Ortiz-Villajos provides an historical explanation for why Spain, as treated by Cebrián and López, had to rely on technology imports during the rapid economic growth of the 1960s. This is the subject of Chapter 8. An international comparison of the number of patent applications revealed that

Spain lagged behind since the beginning of the data in 1820. The author traces this performance to low technological capability, a concept that – like the general level of human capital, social capital, and knowledgeable entrepreneurship – can be subsumed under social capability. Technical education and R&D are the main building blocks of technological capability. For centuries Spain neglected to invest in these areas and thus did not build up the technological capability that would have made it possible to acquire and develop innovations. Consequently the registration of patents, both foreign and domestic, was low. A close analysis of 16,000 patent applications in seven benchmark years between 1882 and 1935 shows that Spanish firms were relatively inactive in this regard, which is regarded as evidence for the low priority accorded to R&D by the private business sector. José Ortiz-Villajos also offers a not too optimistic opinion on recent developments and prospects for the future. Despite Spain's catch-up and increasing wealth, path dependency seems to persist in the area of technological capability.

Technological specialization in Sweden and Finland

Since the mid-nineteenth century, the forestry industry has been of chief importance to the economies of Sweden and Finland. Apart from the recent surge in the telecommunications industry this is the basis of the often exaggerated similarity between the two economies. Sweden was among the successful economies in the Second Industrial Revolution, and many of the innovative 'genius enterprises' established in that period dominated Swedish manufacturing throughout the twentieth century. In Finland, however, employment in the agricultural sector remained above 50 per cent into the 1930s. War reparations, in the form of capital goods to the Soviet Union, became an impetus for the growth of the Finnish engineering industry, and when reparations were completed the market in the east was retained. In the post-war period Finland has progressively been catching up with the rest of western Europe. As regards the growth of GDP per capita, Finland ranked seventh among sixteen countries (the EU 15 minus Luxemburg, but including Norway and Switzerland) in the 'Golden Age' of 1950–73. Contrary to the belief that decline began after 1970, Sweden had already lost steam and was ranked twelfth during the 'Golden Age'. In the 1974–93 period, from the peak just before the structural crisis in energy, steel and shipbuilding, to the trough of the financial EMS crisis,[10] Finland advanced to fifth position. In the same period, the Swedish growth rate sank to fifteenth spot, ahead of only Switzerland. The EMS crisis struck Finland particularly hard, owing to the collapse of the Soviet Union which removed a main export market. In the 1990–3 period Finland decreased her GDP per capita by 12.5 per cent. Sweden was also hard hit, and lost 6.4 per cent over 1991–3. Then came a shift, and between 1994 and 2002 Finnish growth was second only to Ireland and Sweden rose to six in the ranks.[11] Nokia has come to signify the remarkable Finnish performance, even more so than Ericsson has been a recognizable logo for Sweden.

The Finnish and Swedish developments may appear as a story of divergence and convergence. The forestry industry remains strong, and with the successes of Nokia and Ericsson, Sweden and Finland seem to be a rather homogeneous region. In Chapter 9, Svante Lingärde and Jani Saarinen set out to examine whether that picture is accurate. Swedish and Finnish patents, registered in the USA, are compared for the period 1963–97. The reader is introduced to the problem of patents as a measure of technological performance and specialization. Early in the period, the number of Finnish patents in the US amounted to only a fraction of those registered by Swedes and Swedish firms. Yet Finnish patents consistently increased, and approximated the Swedish in the 1990s, on a per capita basis. Like Ortiz-Villajos in Chapter 8, Lingärde and Saarinen calculate the revealed technological advantage (RTA) in order to analyse the sectoral distribution of patents. This indicator is also creatively varied to determine the extent to which patenting correlates with R&D, production and exports. They conclude that, despite Finnish catch-up and the joint elevation of the technological level, heterogeneity rather than homogeneity characterizes the direction of specialization. This finding has a clear bearing on the discussion of the development of contemporary European integration – and broadly is in line with the recent theoretical strand on economics and geography (Fujita, Krugman and Venables 1999). Some new light on issues relating to technology and human capital, in a historical perspective, await the reader who continues to the case studies in the following chapters.

Notes

1 I am indebted to Camilla Josephson, Kyla Madden, Carl-Axel Olsson, Lennart Schön, Jan-Pieter Smits and Lars Svensson, for useful suggestions and comments, but alone take responsibility for any flaws.

2 John Hicks' (1932) idea of technical change 'induced' by relative factor prices may represent the beginning of a comprehensive theory. For discussion see Olsson (1982) and Ruttan (2001, ch. 4)

3 Adam Smith was not explicit on this point but, in a footnote to the third edition, Babbage remarked that the same point 'had been distinctly pointed out' before, by Gioja (1815). Schumpeter (1954 [1982], pp. 511, 541) was harsh on Gioja but connected him with Babbage.

4 Although 'industrial' here is not quite apt since the Enlightenment also encompassed agriculture. An illuminating case is the foundation of Agricultural Societies. With inspiration from England a Danish society was formed in 1769 and a Finnish in 1797. However, the reform of Danish agriculture was already in full progress at the time of foundation of the society, whereas Finnish agriculture had to wait another hundred years for modernization (Christensen 1996, ch. 5; Westerlund 1985).

5 Babbage was a representative of the scientific mentality that characterized the Industrial Enlightenment when he described the role of information, which

also constitutes the feedback mechanism between propositional and prescriptive knowledge: 'it is a maxim equally just in all the arts, and in every science, that *the man who aspires to fortune or to fame by new discoveries, must be content to examine with care the knowledge of his contemporaries, or to exhaust his efforts in inventing again, what he will most probably find has been better executed before'* (Babbage 1833, p. 266; italics in original).

6 On the paradox, see for example David (1990) and Crafts (2001).

7 An interesting development of the GPT approach is provided by Harberger's (1998) division between 'yeast' and 'mushroom' processes of growth. The success or failure of a country to adopt a key technology such as steam, electricity and ICT to a certain extent depends on the existing structure, and on whether it can be applied in broad segments of the economy. This should be revealed in cross-country productivity comparisons, as shown by van Ark and Smits (2004).

8 Mulligan and Sala-i-Martin (1997) estimate the counterfactual level of the zero-schooling wage in the US member states with Mincerian wage equations. Earnings above the zero-schooling level of 'raw labour' represent human capital. The result indicates that human capital is lower in some states with high income levels since the zero-schooling wage is also high. Conversely, human capital is high in low income states such as Arkansas and South Dakota due to very low zero-schooling wages.

9 It is also a risk to be first with the latest, since there has been a substantial development in cointegration analysis and early applications have become obsolete (for a still relevant discussion, see Maddala and Kim 1998).

10 The financial crisis in 1992–93, in the wake of German reunification in 1990 shook the EMS of fixed exchange rates (Ljungberg 2004b).

11 Calculated using Angus Maddison (2001), with the latest update by the Groningen Growth and Development Centre (http://www.ggdc.net).

References

Abramovitz, M. (1956), 'Resource and output trends in the United States since 1870', *American Economic Review*, 46 (1), pp. 5–23, reprinted in M. Abramovitz, *Thinking about Growth and Other Essays on Economic Growth and Welfare* (Cambridge: Cambridge University Press, 1989).

Abramovitz, M. (1979), 'Rapid growth potential and its realization: the experience of capitalist economies in the post-war period', reprinted in Abramovitz (1989).

Abramovitz, M. (1995), 'The Elements of Social Capability', in B. H. Koo and D. H. Perkins (eds), *Social Capability and Long-Term Economic Growth* (London: Macmillan), pp. 19–47.

Angus-Butterworth, L. M. (1958), 'Glass', in C. H. Singer, E. J. Holmyard, A. R. Hall and T. I. Williams, *A History of Technology*, Vol. IV (Oxford: Clarendon Press).

Ark, B. van and J. P. Smits (2004), 'Technology regimes and growth in the Netherlands: an empirical record of two centuries' (mimeo, Groningen Growth and Development Centre).

Babbage, C. (1833), *On the Economy of Machinery and Manufactures*, 3rd edn enlarged (London: Charles Knight).

Berry, W. T. (1958), 'Printing and Related Trades', in C. H. Singer, E. J. Holmyard, A. R. Hall and T. I. Williams, *A History of Technology*, Vol. V (Oxford: Clarendon Press).

Black, J. (1994), 'Continuity and Change in the British Press 1750–1833', *Publishing History*, 36, pp. 39–85.

Black, J. (2001), *The English Press, 1621–1861* (Gloucester: Sutton).

Braverman, H. (1974), *Labor and Monopoly Capital: The Degration of Work in the Twentieth Century* (New York: Monthly Review Press).

Bresnahan, T. F. and M. Trajtenberg (1995), 'General Purpose Technologies: "Engines of Growth"?', *Journal of Econometrics*, 65, pp. 83–108.

Christensen, D. C. (1996), *Det moderne projekt: teknik & kultur i Danmark-Norge 1750–(1814)–1850* (Copenhagen: Gyldendal).

Crafts, N. (2001), 'The Solow Productivity Paradox in Historical Perspective', *CEPR Discussion Paper 3142*.

Crafts, N. F. R. and C. K. Harley (1992), 'Output Growth and the British industrial revolution: a restatement of the Crafts–Harley view', *Economic History Review*, XLV (4), pp. 703–30.

Dahmén, E. (1970), *Entrepreneurial Activity and the Development of Swedish Industry, 1919–1939*, The American Economic Association Translation Series (Homewood, IL: R.D: Irwin; Swedish original 1950).

David, P. A. (1990), 'The Dynamo and the Computer: An Historical Perspective on the Modern Productivity Paradox', *American Economic Review*, 80(2), pp. 355–61.

Denison, E. F. (1962). *The Sources of Economic Growth in the United States and the Alternatives before Us*, Supplementary Paper 13 (New York: Committee for Economic Development).

Doms, M., T. Dunne and K. R. Troske (1997), 'Workers, Wages and Technology', *Quarterly Journal of Economics*, 112 (12), pp. 253–90.

Donovan, A. (1993), 'Education, industry, and the American university', in Fox and Guagnini (1993).

Feinstein, C. H. (1988), 'National Statistics, 1760–1920', in C. H. Feinstein and S. Pollard (eds), *Studies in Capital Formation in the United Kingdom, 1750–1920* (Oxford: Clarendon Press).

Fox, R. and A. Guagnini (eds) (1993), *Education, Technology and Industrial Performance in Europe, 1850–1939* (Cambridge: Cambridge University Press).

Fujita, M., P. Krugman and A. J. Venables (1999), *The Spatial Economy: Cities, Regions and International Trade* (Cambridge, MA: MIT Press)

Gerschenkron, A. (1952), 'Economic Backwardness in Historical Perspective', in B. Hoselitz (ed.), *The Progress of Underdeveloped Countries* (Chicago, IL: Chicago University Press), reprinted in A. Gerschenkron, *Economic Backwardness in Historical Perspective. A Book of Essays* (Cambridge, MA: Belknap Press, 1962).

Gioja, M. (1815), *Nuovo Prospetto delle Scienze Economiche*, vol. 6, Milan.

Goldin, C. and L. F. Katz (1998), 'The Origins of Technology–Skill Complementarity', *Quarterly Journal of Economics*, 113 (3), pp. 693–732.

Harberger, A. C. (1998), 'A Vision of the Growth Process', *American Economic Review*, 88 (1), pp. 1–32.

Harley, C. K. (1991), 'Substitution for prerequisites: endogenous institutions and comparative economic history', in R. Sylla and G. Toniolo, *Patterns of European Industrialization. The Nineteenth Century* (London: Routledge).

Harley, C. K. (1993), 'Reassessing the Industrial Revolution: A Macro View', in J. Mokyr (ed.), *The British Industrial Revolution. An Economic Perspective* (Boulder, CO: Westview Press).

Harley, C. K. and N. F. R. Crafts (2000), 'Simulating the Two Views of the British Industrial Revolution', *Journal of Economic History*, 60 (3), pp. 819–41.

Headrick, D. R. (2000), *When Information Came of Age. Technologies of Knowledge in the Age of Reason and Revolution, 1700–1850* (Oxford: Oxford University Press).

Helpman, E. (ed.) (1998), *General Purpose Technologies and Economic Growth* (Cambridge, MA: MIT Press).

Hicks, J. R. (1932), *The Theory of Wages* (London: Macmillan; 2nd edn 1965).

Jorgenson, D. W. (1966), 'The Embodiment Hypothesis', *Journal of Political Economy*, 74 (1), pp. 1–17, reprinted in D. W. Jorgenson, *Productivity. Volume 1: Post-war U.S. Economic Growth* (Cambridge, MA: MIT Press, 1995).

Jorgenson, D. W. (2001) 'Information technology and the U.S. Economy', *American Economic Review*, 91 (1), pp. 1–32.

Jorgenson, D. W. and Z. Griliches (1967), 'The Explanation of Productivity Change', *Review of Economic Studies*, 34 (3), pp. 249–80, reprinted in Jorgenson (1995).

Katz, L. F. and K. M. Murphy (1992), 'Changes in Relative Wages, 1963–1987: Supply and Demand Factors', *Quarterly Journal of Economics*, 107 (1), pp. 35–78.

Kuznets, S. (1966), *Economic Growth and Structure. Selected Essays* (London: Heinemann Educational Books).

Landes, D. (1998), *The Wealth and Poverty of Nations. Why Some Are So Rich and Some So Poor* (New York: Norton).

Lindert, P. H. and J. G. Williamson (2001), 'Does globalization make the world more unequal?', *NBER Working Paper 8228*.

Ljungberg, J. (1990), *Priser och marknadskrafter i Sverige 1885–1969. En prishistorisk studie* (Lund: Ekonomisk-historiska föreningen).

Ljungberg, J. (1991), 'Prices and Industrial Transformation', *Scandinavian Economic History Review*, XXXIX (2), pp. 49–63.

Ljungberg, J. (1996), 'Catch-Up and Static Equilibrium: A Critique of the Convergence Model', *Scandinavian Economic History Review*, XLIV (3), pp. 265–75.

Ljungberg, J. (1997), 'The Impact of the Great Emigration on the Swedish Economy', *Scandinavian Economic History Review*, XLV (2), pp. 159–89.

Ljungberg, J. (2004a), 'Earnings differentials and productivity in Sweden, 1870–1980', in S. Heikkinen and J. L. van Zanden (eds), *Exploring Economic Growth* (Amsterdam: International Institute of Social History).

Ljungberg, J. (2004b), 'The EMU in a European Perspective. Lessons from monetary regimes in the 20th century', in J. Ljungberg (ed.), *The Price of the Euro* (Basingstoke: Palgrave Macmillan).

Lucas, R. E. Jr (1993), 'Making a Miracle', *Econometrica* 61(2), pp. 251–72.

Maddala, G. S. and I.-M. Kim (1998), *Unit Roots, Cointegration and Structural Change* (Cambridge: Cambridge University Press).

Maddison, A. (1987), 'Growth and Slowdown in Advanced Capitalist Economies: Techniques of Quantitative Assessment', *Journal of Economic Literature*, XXV, pp. 649–98.

Maddison, A. (1995), *Monitoring the World Economy, 1820–1992* (Paris: OECD).

Maddison, A. (2001), *The World Economy. A Millennial Perspective* (Paris: OECD).

Mitch, D. (1993), 'The Role of Human Capital in the First Industrial Revolution', in J. Mokyr (ed.), *The British Industrial Revolution. An Economic Perspective* (Boulder, CO: Westview Press).

Mokyr, J. (2000), 'Knowledge, Technology, and Economic Growth during the Industrial Revolution', in B. van Ark, S. K. Kuipers and G. H. Kuper, *Productivity, Technology and Economic Growth* (Dordrecht: Kluwer Academic).

Mokyr, J. (2002), *The Gifts of Athena. Historical Origins of the Knowledge Economy* (Princeton, NJ: Princeton University Press).

Mulligan, C. B. and X. Sala-i-Martin (1997), 'A labor income-based measure of the value of human capital: An application to the states of the United States', *Japan and the World Economy*, 9, pp. 159–91.

Nilsson, A., L. Pettersson and P. Svensson (1999), 'Agrarian transition and literacy: the case of nineteenth-century Sweden', *European Review of Economic History*, 3 (1), pp. 79–96.

Olsson, C.-A. (1982), 'Relative Factor Prices and Technical Change: Some Theoretical and Historical Perspectives', in L. Jörberg and N. Rosenberg, *Technical Change, Employment and Investment*, Theme A3 at the Eighth International Economic History Congress, Budapest 1982 (Lund: Department of Economic History).

O'Rourke, K. H. and J. G. Williamson (1995a), 'Open Economy Forces and Late Nineteenth Century Swedish Catch-Up. A Quantitative Accounting', *Scandinavian Economic History Review*, XLIII (2), pp. 171–203.

O'Rourke, K. H. and J. G. Williamson (1995b), 'Education, Globalization and Catch-Up: Scandinavia in the Swedish Mirror', *Scandinavian Economic History Review*, XLIII (3), pp. 287–309.

O'Rourke, K. H. and J. G. Williamson (1999), *Globalization and History. The Evolution of a Nineteenth-Century Atlantic Economy* (Cambridge, MA: MIT Press).

Petersson, K. M., A. Reis and M. Ingvar (2001), 'Cognitive processing in literate and illiterate subjects: a review of some behavioral and functional neuroimaging data', *Scandinavian Journal of Psychology*, 42 (3), pp. 251–67.

Pollard, S. (1981), *Peaceful Conquest. The Industrialization of Europe, 1760–1970* (Oxford: Oxford University Press).

Prados de la Escosura, L. (2003), *El progreso económico de España, 1850–2000* (Madrid: Fundación BBVA).

Romer, P. M. (1990), 'Endogenous Technological Change', *Journal of Political Economy*, 98 (5), pp. S71–S102.

Rosenberg, N. (1982), *Inside the Black Box: Technology and Economics* (Cambridge: Cambridge University Press).

Rosenberg, N. (1994), *Exploring the Black Box: Technology, Economics, and History* (Cambridge: Cambridge University Press).

Ruttan, V. W. (2001), *Technology Growth and Development. An Induced Innovation Perspective* (New York: Oxford University Press).

Schumpeter, J. A. (1954), *History of Economic Analysis* (reprinted London: Allen & Unwin, 1982).

Schön, L. (1989), *From War Economy to State Debt Policy* (Stockholm: Riksgäldskontoret).

Schön, L. (1991), 'Development blocks and transformation pressure in a macroeconomic perspective – a model of long-term cyclical change', *Skandinaviska Enskilda Banken Quarterly Review*, 20 (3–4), pp. 67–76.

Schön, L. (2000), *En modern svensk ekonomisk historia. Tillväxt och omvandling under två sekel* (Stockholm: SNS).

SHNA: Swedish Historical National Accounts (Lund: Ekonomisk historiska föreningen), in nine volumes:

Krantz, O. (1986), *Offentlig verksamhet 1800–1980*.

Krantz, O. (1987a), *Transporter och kommunikationer 1800–1980*.

Krantz, O. (1987b), *Husligt arbete 1800–1980*.

Krantz, O. (1991), *Privata tjänster och bostadsutnyttjande 1800–1980*.

Ljungberg, J. (1988), *Deflatorer för industriproduktionen 1888–1955*.

Pettersson, L. (1987), *Byggnads– och anläggningsverksamhet 1800–1980*.

Schön, L. (1984), *Utrikeshandel 1800–1980*. Mimeo.

Schön, L. (1988), *Industri och hantverk 1800–1980*.

Schön, L. (1995), *Jordbruk med binäringar 1800–1980*.

Svensson, P. (2001), *Agrara entreprenörer. Böndernas roll i omvandlingen av jordbruket i Skåne ca 1800–1870* (Lund: Lund Studies in Economic History), 16.

Tortella, G. (1994), 'Patterns of economic retardation in south-western Europe in the nineteenth and twentieth centuries', *Economic History Review*, XLVII, pp. 1–21.

Westerlund, L. (1985), *Hushållnings- och lantbrukssällskapen I Finland åren 1797–1909 – plattformar i länen för samhälleligt deltagande*, Meddelanden från ekonomisk-statsvetenskapliga fakulteten vid Åbo Akademi (Åbo: Åbo Akademis kopieringscentral).

Williamson, J. G. (2000), 'Globalization and the labor market: using history to inform policy', in P. Aghion and J. G. Williamson, *Growth, Inequality and Globalization* (Cambridge: Cambridge University Press).

2
Human Capital and Industrialization: The Case of a Latecomer – Portugal, 1890[1]

Jaime Reis

Introduction

This chapter touches on two inter-related debates in contemporary economic history concerning the role of human capital in economic growth. The first is the question of whether the early stages of the industrialization process were characterized essentially by a substitution of capital and unskilled labour, for skilled labour and therefore by a low degree of complementarity between technology (usually proxied in the literature by fixed capital) and human capital. The conventional view is that it was indeed a de-skilling process and that there was a high degree of substitutability between physical and human capital. The second debate arises from the consensus among growth economists that human capital is an important factor of growth in the long run and has been responsible for a good deal of convergence between economies (Abramovitz 1993). However, given the concurrent view about the limited importance of human capital in the Industrial Revolution, this has led many to accept the notion that this could not have happened during the nineteenth century but only during the twentieth century. Goldin and Katz (1998; 2000) and Goldin (2001) have tried to give substance to this argument with the claim that the reason for this secular transition has to do with the evolving nature of technology. In their view, the rise of capital – skill complementarity only in the twentieth century was determined by the spread of batch and continuous process production, the diffusion of electrical power and the rise of industries that were deeply permeated by scientific applications, all of which have only occurred in the last one hundred years.

The study proposed here is in line with the recent revisionist work which diverges from the assumptions that underlie this mainstream view (Boot 1995; Rosés 1998; Bessen 2003). It considers that during the nineteenth century the most important form of human capital for industry was the skills relevant to the job being done. These might comprise literacy and other forms of formal education and be acquired by children and young people through schooling; but they also included practical know-how and experience achieved in the workplace. The latter represented an investment that enhanced workers' productive capacity and can be measured by the price fetched in the market by the flow of services it generated. It is revealed by the gap between the wages of raw and those of skilled labour. During training, such trainees may even have earned less than the unskilled given that their acquisition of know-how had a cost, but accumulated over their lifetime they came to earn much more. Second, descriptions of factory work suggest that the more intense division of labour that accompanied industrialization did not preclude the need for skills but created a demand for skills that were different from, though not inferior to, those in craft-based production. These new skills became necessary in large quantities long before the era of electricity and batch production. In the early factories, greatly increased throughput and capital intensity compelled operatives to work faster, with greater regularity, and to avoid mistakes that might cause costly, capital-intensive stoppages or damage to equipment. Thus technological progress required not only increased skills for the maintenance of machinery but also increased skills in many other areas of production.

. To date, too little empirical evidence has been presented on these matters and most of it has had to do with the advanced economies of the twentieth century. This chapter, which studies the case of Portugal in the late eighteen hundreds, seeks to broaden this rather limited geographic perspective regarding the role of skill in industrialization during the 'first Industrial Revolution'. In the second place, its focus on a late developing economy from the periphery of Europe allows for consideration of certain interesting features of such economies. One is their status as importers rather than creators of new technology, which could lead to the use of equipment that might be more suitable for a different factor mix. Another is their lack of industrial background and the consequent scarcity of human capital appropriate to rapid industrial development.

During the second half of the nineteenth century, Portugal was far from being an industrialized economy. In 1852, manufacturing establishments with more than ten workers had a work force of just under 16,000 individuals for a total population of close to 4 million (*Relatório* 1857). Following several decades of moderate growth of industrial output, at an average rate of 2.8 per cent per annum, industrial labour in 1890 was still only 18 per cent of the total labour force, whereas agriculture was 61 per cent and agricultural output in 1900 was something like twice the size of that of industry (Nunes 1989). Factories were concentrated in and around the main coastal cities, namely Lisbon and Porto, with the exception of Covilhã, in the mountainous interior, which had been a traditional centre for woollens since the eighteenth century. The structure of the manufacturing sector was typical of economies which embarked late and slowly on the process of industrialization, had a scarcity of suitable raw materials and a heavily protected domestic market. Around 1910, it was dominated by cottons, woollens, metalworking and food, followed at a distance by cork, fish canning, ceramics and tobacco (Lains 2003). Low productivity in all branches – somewhere around half of that in comparable situations in Britain and France – practically disqualified Portuguese manufacturers from competing internationally and obliged industry to shelter behind high tariffs. The exceptions were cork and fish canning, two industries whose advantage lay in an unusual local abundance of highly specific raw materials: the bark of cork trees and sardines. In general, technological progress appears to have been slow, significant mechanization bypassed many firms, and the use of steam and waterpower was not impressive and neither was it always successful. Craft and domestic production therefore managed to adapt and survive to a remarkable extent (Pedreira 1990; Pereira 2001).

At the end of the nineteenth century, the state carried out two major national inquiries, or *Inquéritos*, into the condition of Portuguese industry, respectively in 1881 and 1890.[2] Both met with mixed results largely as a result of resistance on the part of entrepreneurs. Information was gathered from a fraction only of the firms in question and often the questionnaires were incompletely answered. They were concerned with a wide range of aspects, such as the value of fixed and circulating capital, raw materials and output; the number, gender, age, wages and literacy of the work force; and the sources of power and the machinery employed. Apart from these statistics, numerous interviews and visits to factories were undertaken and provide additional valuable insights.

The present study makes use principally of the second of these inquiries because it is the most informative in terms of the remuneration of labour.[3] We have chosen to focus on Lisbon for two reasons. It was much the largest industrial centre of the time (Porto, in second place, had half the number of industrial workers). In addition, it had a much greater share of steam powered factory-based production, which was where technological progress was likely to occur (Cordeiro 1996; Pereira 2001). Altogether, in the capital, 2,900 questionnaires were sent out to as many firms, which gives an idea of the size of the universe in question. Only 1,614 were returned and they covered firms employing a total of 17,403 men, women and children, but the total work force in Lisbon's manufacturing can be reckoned, by extrapolation, at approximately 31,000.[4] Since our interest lies in the link between technological and skill development, artisanal units are left out of our sample, meaning that only firms with more than ten workers are therefore considered.

This chapter consists of four main sections. The first one shows that skills were important for Portugal's late nineteenth-century industrial development. The Portuguese manufacturing labour force contained a large proportion of skilled workers (that is, those who earned skill premia), and their distribution is discussed in terms of the technological sophistication of the branches of manufacturing to which they belonged. The second part tries to ascertain whether literacy contributed to the working skills used on the shop floor. It checks on what impact cognitive skills had in this respect and whether it was practical learning prior to adulthood which was the most important factor in the development of industrial human capital. The third part tries to gauge the importance of this stock of human capital in industrial production. This is done in a growth accounting framework that estimates the contributions to value added of capital, raw labour and human capital. The fourth and final section tests for the presence of capital – skill complementarity in Portuguese industry at this time, well before it possessed the characteristics which have been claimed in the literature as essential for this. It uses skill premia rather than the educational achievement of the work force as the yardstick for human capital and finds that such complementarity did indeed exist.

A skilled industrial labour force

In this part of the chapter, we try to measure the extent to which human capital was present in the Lisbon industrial work force in 1890. Human capital here is equated with job-related skills, and is proxied by the market value of the services derived from this endowment. As proposed by McIvor (2001, p. 48), it was 'a combination of manual dexterity with knowledge (of materials and tools) and discretion acquired through a long training period, traditionally of several years'. It is measured here by the premium paid to each worker above the remuneration of raw labour, the latter corresponding to the effort of workers who were totally unskilled.

Several doubts can be raised regarding such a Mincerian approach (Mincer 1974). Besides differences in human capital endowment, individual wage gaps could be explained by variations in innate ability and strength among workers. They could also stem from imperfections in the labour market that would make wages less likely to equal the marginal product of labour. The first type of distortion is automatically avoided with the 1890 *Inquérito* because the latter does not allow the use of individual observations anyway. It registers only the maximum and minimum daily wage for the workers in a given job category in each productive unit (for example, carpenters in firm X). The information contained both in the 1881 *Inquérito* and in the trade union inquiry of 1909 (Cabral 1977) tells us that the averages of these maxima and minima provide a good approximation to the average wage earned in each of these occupational categories.[5] This is the variable we shall be using here, the advantage being that we can thus reasonably ignore the difficulty posed by differences in individual characteristics. The disadvantage is that while there are large pay discrepancies associated with gender, the 1890 data fail to make any distinction in remuneration between males and females. In mixed groups, presumably the 'best' men earned the maximum wage and the 'worst' women received the minimum but we do not know what were their respective minima and maxima, or, consequently, the average for each gender in that occupation. This forces us to restrict ourselves to firms that are homogeneous in gender terms and thus all firms employing both males and females are excluded here from observation.

A second distortion could be caused either by factors such as employer paternalism or discrimination regarding certain groups, or by the payment of 'efficiency wages'. The latter could take place as a disincentive to shirking and other forms of opportunistic behaviour. They could assume the form of payments for assiduity and 'good behaviour' coming towards the end of the worker's career. In the short run, this could also be done to hold on to skilled workers during a cyclical downturn, in the expectation of an upturn (Filer, Hamermesh and Rees 1996). In the cases of nineteenth-century Lancashire and Catalonia, Boot (1995) and Rosés (1998), respectively, have shown that there was little reason for concern about these effects and the same position seems warranted in the present instance too. For one thing, the evidence on age–wage profiles for Lisbon, although scant, shows a clear decline in wages at the end of the working lives of industrial workers, not a rise, as the argument would require (Carqueja 1916). This is reinforced by the evidence from trade unions at the beginning of the twentieth century, which maintained that employers only kept workers on the payroll as long as they were healthy, productive and necessary (Cabral 1977).[6] The same source stressed that managers paid wages in accordance with what workers produced, even when not on a piece rate system, a view that contradicts the existence either of 'paternalism' or of 'efficiency wages'.[7] Finally, monitoring shirking should not have been the problem it was in other historical contexts. Most Lisbon firms were neither large nor adopting new, cutting-edge technology, both of which were the circumstances that made it difficult for employers to know how much workers were really able to produce (Huberman 1986; 1991).

The assumption of a freely competitive and transparent labour market may have been violated by two further conditions. One is that the tariff protection enjoyed by manufacturing was hardly uniform across the economy. Wage differences for the same task between firms or sectors could have arisen because some firms would have been able to pay more out of the economic rents that accrued from the higher protection they received independently of productivity differences. The second distortion could have been generated by the exceptional market power regarding worker recruitment that some larger firms might have enjoyed compared to smaller ones. As regards the latter, the problem was probably not serious for two reasons. Truly large firms, with several hundreds of operatives, were not common in Lisbon at this time and did not dominate any branch of industry.[8] In the second place, although some occupations (such as spinning and weaving) were branch-specific, in most cases jobs, such as carpenters or fitters, were common to a number of industrial branches and the market for them was therefore quite extensive and presumably competitive. It is more difficult to allay worries regarding the former distortion since there are in fact signs of a marked lack of inter-sectoral uniformity when it comes to tariffs. Nominal *ad valorem* protection ranged from 14.6 to 81.5 per cent (Lains 2003, p. 118). Little is known, however, about effective (as opposed to nominal) protection rates, and without this knowledge it is impossible to draw any conclusions regarding the impact of this on the issue at hand. In any case, problems would only arise if we assumed that competition among firms for certain skills was weak, and this (as we have seen) may not have been the case.

A final problem is raised by Gregory Clark's (1994) claim that, during the Industrial Revolution, the wage of factory workers included a premium for their submission to this harsh and novel form of discipline. This rewarded them for working longer hours and also harder, so that they produced and earned more per hour than their counterparts in the non-factory sector. As regards the assumption we are following here, namely that wages were close to marginal productivity, it is not clear that Clark's analysis invalidates it. The principal reason is that the wage of raw labour pertained to the same working environment as that of the skilled workers under examination: that is, the factory. If there was a 'discipline premium', presumably it had to be paid to both categories of workers, and therefore the difference between their respective wages would have equalled the 'true' skill premium which we are using to proxy human capital.[9] A further reason is that Clark's demonstration that harsher discipline required extra payment is based on the notion that factory and non-factory workers had the same skills. This is far from certain. If, on the contrary, factory workers had been more skilled, their higher wages might simply reflect this, and then a 'discipline premium' might not have existed at all. In any case, being able to accept factory discipline was not just a matter of pay but also of a socialization that prepared workers for it. Whether this was learned at school or in the factory, our method of estimation of human capital would still have been valid because workers who put up with these conditions had a personal attribute that contributed to productivity and were remunerated accordingly.

The rates of pay considered here are hourly, not daily. This is important because there was quite a range of lengths of working day in 1890.[10] The 1890 *Inquérito* distinguishes in addition between the length of summer and winter working days, the latter being shorter by one to two hours. We incorporate this information by assuming for the whole year an average of these two figures. In cases of firms where complete information was unobtainable, we presumed that they followed the timetable of the other businesses in the same branch. Where a range of hours was given, instead of a single figure, we took the mean of the extremes of the range. Children have been left out of this analysis because the data available for young workers under sixteen is in a range of variation that is strongly affected by age and physical strength, besides skill. This renders it difficult to establish how much of any wage gap is ascribable to skill alone. In the case of women, the number who satisfied the requirement of belonging to an occupational group that did not incorporate any men was too small – 333 in Lisbon – to be meaningful. Altogether, we have been able to take into account 6,135 adult male workers divided by occupation and firm into 142 different groups. The average hourly rates for these men varied enormously, between 236 and 20 reis.[11]

The data on wages are summarized in Table 2.1. The first point to note is that the lowest paid category was of those receiving 40 reis or less per hour (the normal rate for unskilled workers in Lisbon) and corresponded to the greatest level of physical effort, without any skill.[12] According to our tabulation, just over 4 per cent of the male adult labour force belonged in this group and this means that the overwhelming majority of industrial operatives performed tasks that involved some skill, however acquired. The well-off workers, at the other end of the spectrum, were those who received more than 100 reis an hour. They can be compared, in terms of earning power, to minor civil servants and could comfortably support a family of four or five on one wage alone.[13] This group comprised only 697 individuals, or just over 11 per cent of the sample considered. Immediately below them were another 2,086 males earning double or more the wage of unskilled workers and therefore still at an immense economic distance from the latter. Between these two strata and that of raw labour were the remaining 50 per cent of this work force (that is, just over 3,000 employees), who would face hardship if they had to live on the earnings of a single bread-winner or if there was temporary unemployment (Poinsard 1910). Altogether, the picture that emerges is that of a surprisingly qualified work force, where premia for skills rewarded a large proportion of its members, and few belonged in the stratum at the bottom of the earnings heap where workers were paid exclusively for their strength and stamina.

To what kinds of occupations and branches of manufacturing did these classes of remuneration correspond? Following the organization provided by Table 2.1, the first thing to note is that in the top group, of those earning over

Table 2.1 Distribution of hourly earnings of adult male industrial workers*

Wages	Number >16 years	Percentage
<41 reis	274	4.5
41–60 reis	1,059	17.3
61–80 reis	2,019	32.9
81–100	2,086	34.0
>100 reis	697	11.3
Total	6,135	100

* Wage is the average, for each occupational group in each firm, of the maximum and minimum hourly pay.
Source: Inquérito Industrial (1890).

100 reis an hour, jobs were characterized by considerable manual dexterity, even some degree of artistic capacity, and the operation of sophisticated precision machinery, such as printing presses or equipment for minting, electroplating and zincography. Some of these activities involved recent technological developments based on electrical and/or chemical processes. In some cases – for example, M. Herman, a manufacturer of electrical apparatuses – worker categories were not even described but only the product, probably because their jobs defied traditional classifications. Some occupations (such as carpenters, metal workers, boiler-makers and blacksmiths), which were usually situated further down the hierarchy of pay, intruded into this elite apparently when they became engaged in custom-made production.

The category immediately below, of those earning from 81 to 100 reis per hour, contains a very wide variety of skilled occupations, many of them to be found also in the previous group, such as typographers, compositors and master dyers, though less prevalent than there. New and significant presences concern tasks relating to the operation or production of steam engines, such as machine operators and boiler makers. A few carpenters, stonemasons, painters and others associated with construction also appear now along with less skilled members of the precision trades of the preceding category. The salient feature of this group is its dominance by trades connected with metal, including blacksmiths, boilermakers, turners, founders and gun makers. Naval construction appears here too, with a large contingent of carpenters who specialized in shipbuilding, an occupation with close links to some of the metal trades. Other trades associated with wood, such as sawyers, begin to emerge in the lower ranks of this earnings category, as happens also with the modern segment of the food industry, such as steam-powered flour milling. Textiles are represented by their most skilled workers (dyers and finishers).

The third group, earning from 41 to 80 reis, encompasses a much broader range of occupations and industries. It still includes a representation of metal and precision workers but now in quite small numbers. The bulk comes from the 'traditional' branches of industry: that is, food (sugar, bread, beer and margarine, but not meat because the abattoir was municipalized and had artificially high wages), leather and shoes, rope, soap, furniture, construction and the production of its inputs (nails, sawing, quick lime). Most of them were consumer-oriented sectors, where technical progress, including mechanization, had been relatively scarce, long-term growth had been slow and their origins went further back in time.[14]

In the last and smallest category of Table 2.1, consisting of unskilled, low paid males, one finds hardly any of the jobs listed in the top first and second categories. On the other hand, many have no specific description at all and belong to vague occupations such as 'worker', 'operative', 'attendant' or 'helper'.[15] To these must be added all whose tasks had little specialized content, such as guards, carters, drivers and doormen. Low skill occupations such as washers, starchers and bakers are also present.

It thus would seem that in 1890, among male adult workers, wage gaps were very much related to the exigencies of the tasks they had to carry out. Those that did the more demanding jobs in terms of skill on the whole were higher up the pay scale, while those whose jobs required mostly brute force tended to be at the other end of it. As theory would have it, employers seem to have valued their workers' skills or human capital by paying them a premium over and above what was earned by the workers who lacked these attributes; in other words, unskilled workers. The former comprised a predominant share of the Portuguese industrial labour force at this time.

The nature of human capital

So far we have not examined the exact nature of the human capital considered here or the manner of its acquisition. Broadly speaking, the literature considers two possibilities in this respect. The most common equates this factor of production with aptitudes of a cognitive nature that are acquired in the course of formal education, mainly carried out in schools. The indicators used in such studies have been those relating to scholastic performance: for example, rates of literacy or of success in exams, school enrolment or attendance. Measures such as the number of years of schooling per individual or the proportion of the population of school age that is registered in educational establishments are also commonly related to the attribute we are trying to assess. On the other hand, it is clear from many accounts that labour productivity in nineteenth-century manufacturing depended crucially on job-specific skills and that these had to be learned by young workers on the shop floor. This does not mean, however, that prior or concurrent schooling of workers of whatever age did not affect productivity too. Literacy could have enabled the process of learning skills on the job to be faster and more efficient and it could have made it easier for workers to be more receptive to new technologies. It could have made skilled workers that had this capability more effective in their tasks than illiterate ones, as a result of a complementarity between skill and literacy, as Bessen (2003) found in Lowell during the first half of the nineteenth century. Finally, the socialization process undergone by children in the school may have been of such a nature and intensity as to prepare them, once adults, to act more efficiently within the context of large scale, mechanized production, because it rendered them more docile, more punctual and more adaptive to routine.

Contrary claims have also been made for the notion that literacy, numeracy or other cognitive skills had scarce impact on industrial labour's productivity because they had little to do with the acquisition and use of directly productive skills. According to Mitch (1992; 1999), the impact of literacy on economic performance during the early industrialization of Britain was very limited and, consequently, this country may even have greatly over-invested in education and implicitly squandered the resources thus employed. This contrasts with the mainstream view, which rarely considers in depth what the precise effect of the cognitive or the non-cognitive aptitudes learned in school was on workers' productive capacities, and takes for granted their positive and significant contribution to the specific skills required by their jobs (Matthews, Feinstein and Odling-Smee 1982; Maddison, 1995; Crafts, 1995).[16]

In the Portuguese case, opinions have divided along similar lines although the issue has received far less attention. Using qualitative sources, both Reis (1986) and Rodrigues and Mendes (1999) have argued that human capital shortages were a major stumbling block for Portuguese nineteenth-century industrialization. This is based mainly on declarations made by employers and trade unionists in the course of the industrial inquiries referred to above. Mónica (1987), on the other hand, has been sceptical of this on the grounds of lack of convincing evidence that education made workers more productive. One suggestive fact is that industrial operatives certainly achieved above-average educational standards for the period and, simultaneously, were paid above what most other workers earned. Among adult male industrial workers in Lisbon, the literate were close to 63 per cent of the total, while the rate for the population at large was 27 per cent. The majority of non-manufacturing workers, on the other hand, earned a fraction of their pay.

Although interesting, this is hardly conclusive regarding the causality that we are seeking to explore. In order to shed further light on the matter, two approaches can be followed. One requires establishing a relation between indicators of the human capital of individual workers and their respective output. Unfortunately, such studies are difficult to carry out and therefore rare. An important one is that by Bessen (2003), using detailed business records. In it, he has been able to show that literate workers were indeed quicker at learning factory tasks and, once they had done so, they used the skills thus acquired more effectively than their illiterate counterparts, with the result that they were more productive and earned more. The second approach, which is adopted here, considers pay rather than output, treats workers in groups rather than as individuals, and proxies human capital as the excess of workers' average remuneration over that of raw labour.

This second methodology is less satisfactory because it requires that markets be efficient, but it is the only feasible one with the available data. The latter, as seen above, allows us to measure the skill premia by which we proxy the average human capital of occupational groups according to the job-specific-skill view that we are following here. This is the dependent variable in a model which tries to show what were the main determinants of the accumulation of human capital in the labour force. The conjecture is that, when workers started working as adults, in the over-sixteen age group singled out by the 1890 *Inquérito*, they could bring with them two basic aptitudes. One was literacy, attained as a result of schooling; the other was skills and experience acquired on the shop floor while working as children, as teenagers or as young adults.[17] Plausibly, either of these was relevant to the development of the future job-specific skills that would enable them to progress from minimal wage rates, at the start of their careers, to the higher rates of pay corresponding to the full development of their capacities, presumably in their twenties or thirties. These are the explanatory variables that we shall consider.

Contemporary testimony confirms that both suppositions are conceivable. As regards on-the-job skills, several sources indicate that workers under the age of 16 were able to improve their wages year by year, and that this was the result not only of their increasing strength but also of the stock of human capital they built up through work experience. The factory inspector's report for the districts of Évora, Beja and Faro, in 1907, which provides detailed information on the pay of workers at different ages in a variety of branches, commented that growing 'organic forces' (that is, strength and size) were partly responsible for the upward slope of the wage–age profile. But it pointed out that these pay rises were also related to the skills that they had learned on the job. Figure 2.1 presents the wage–age profiles of six occupations graphically, although there are many more in the source that could be used. It makes it plain that as youths entered 'working adulthood', the human capital they had already built up could vary immensely.[18] Some, like tin sheet and fish canning workers, had gone far by this stage, while others, such as most cork workers or spinners, had barely improved their skills. The latter were described as 'beginners' who carried out no more than 'elementary operations'. The hypothesis this suggests for testing is that this initial endowment influenced the future course of practical learning, thereby determining in part the pay differentials encountered among mature workers as a result of their skills.

At the same time, well-informed contemporaries accepted that literacy could also contribute to the productivity of labour. Several employers told the organizers of the 1881 industrial inquiry that one of their problems with the labour force was its lack of 'general education' and linked this to its inefficiency (Reis 1986). It is interesting to note that all the same they tended to recruit illiterate boys and girls when they entered the factory, typically at around eleven or twelve years of age. It is equally remarkable that quite a number of these youths later went on to night school and as adults came to be registered as literate. This is suggested by the fact that the literacy rate of adult factory operatives was generally higher than that of their juvenile counterparts. In some factories, managers said that the better-paid jobs for adults were unavailable to illiterates, which may account for this late educational effort and suggests that literacy might indeed be part of the human capital relevant to industrial productivity.[19]

The 1890 *Inquérito* provides quantitative evidence on both of these explanatory factors. It contains the minimum wage rate for every occupational group, firm by firm, for adult workers. We assume that this was the pay of the youngest and least experienced adults **[minw]** and that it proxies

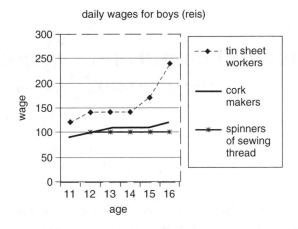

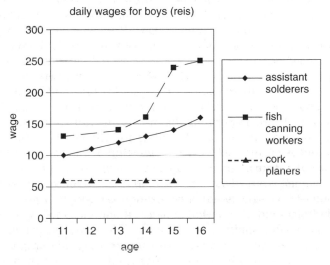

Figure 2.1 Wage–age profiles for boys (reis per day)

Note: data from the districts of Beja, Évora and Faro.

Source: Boletim do Trabalho Industrial (1907), No. 3.

the possession of practical skills by individuals entering the adult work force, therefore before their learning curve began to rise significantly. For each of these groups the *Inquérito* also gives us their average literacy rate as measured by the ability to read only, a figure that at this time is close, however, to that for full literacy.[20] We employ this as a proxy for the stock of cognitive skills of the adult male labour force **[lit]**. The principal shortcoming of both classes of data is that they are not individualized. They refer instead to groups of operatives in the same jobs within firms, but we have no alternative to treating them as individual observations.

In the estimation below, we have included two other variables which may also be pertinent. One is the ratio of the average literacy of adults in a given occupation and firm to that of its juvenile workers (under sixteen) in the same category **[litratio]**. This helps to establish if adult workers in a given job tended to become literate after they grew up in order to secure the better-paid jobs that went preferentially to the literate. The second **[interact]** is an interaction term which serves to test whether workers who were literate were thereby enabled to use their practical skills better than their illiterate colleagues. It is given by the product of the adult literacy rate and the adult male wage.

The results of the ordinary least squares (OLS) treatment of this evidence are displayed in Table 2.2. Two main conclusions can be drawn from it. In the first place, adult literacy did have some effect on the determination of wages by increasing the premium relative to raw labour earned by skilled industrial workers (equation 1). The coefficient on this variable has the correct positive sign and is statistically significant.[21] In other words, at this stage of technological development, the human capital package that employers sought in their workers included skills of the kind imparted by formal schooling, as most of the international literature has claimed. It should be noted, however, that this influence was very weak: the value of R^2 is quite low when the wage premium is regressed on literacy alone.

Table 2.2 The determinants of the skill premia of adult males

Dependent variable: Hkcap

(ordinary least squares regression)

Variables/Equation	1	2	3	4
C	35.57	−6.04	−5.80	18.88
	(10.76)	(−1.08)	(−0.74)	(2.60)
Lit > 16	18.62	4.03	−	−62.02
	(3.92)	(1.68)		−(−5.70)
Minw	−	0.88	1.00	0.38
		(11.38)	(6.92)	(3.16)
Interact	−	−	−	0.10
				(6.48)
Litratio	−	−	0.28	−
			(0.164)	
Adjusted R^2	0.06	0.54	0.57	0.78
F-statistic	19.69	162.35	55.70	337.04
N	285	285	188	285

Note: T-statistics in brackets.

All estimations passed the White heteroscedasticity test satisfactorily.

Our second conclusion is that job-specific skills obtained before adulthood not only made a difference too but they made a much greater one in terms of the degree of human capital acquired on the shop floor during adulthood (equation 2). The coefficient for the minimum adult wage **[minw]** variable has a positive sign and is highly significant. Moreover, it increases the size of R^2 substantially when added to the preceding regression. This suggests a path dependence in the formation of human capital. Youth capabilities were a good predictor for how far a worker could go in his career, and individuals that got the best jobs tended to be selected and trained for them early in life. How this selection was done cannot be determined with the help of the present data set and we can only speculate regarding the probability that family links, social networks and so on may have played a role. In any case, the impression is that skill premia for most workers after the age of sixteen were linked to previous accumulation of experience and practical knowledge on the job. As in England during the period 1780–1860, 'of far greater importance [than formal education] to the growth of a suitable labour force was the training in new, practical skills disseminated in the home, the workshop and the factory, a training which developed along with the rise of the industrial economy rather than preceded it' (Tranter 1981, p. 224).

The two other variables **[interact** and **litratio]** do not improve the model's explanatory power. Introducing the literacy ratio **[litratio]** barely raises the value of R^2 but it makes both the literacy variable **[lit]** and the new variable lose their significance altogether (hence the absence of **[lit]** from the table). Introducing the interaction term **[interact]** improves R^2 by quite a lot but it also changes the sign of the literacy variable **[lit]** to negative, which totally upsets the sense of our analysis.

This leads to two further conclusions. One is that being literate apparently did not enable workers to get more out of their on-the-job skills in terms of productivity and pay premia. The other is that, contrary to the claims encountered in the 1881 *Inquérito*, adult workers were not acquiring reading and writing skills in order to position themselves for better-paid work opportunities. The high correlation between youth and adult literacy rates by occupational group ($r = 0.63$) points instead to a fairly strong lifetime path dependency as regards this attribute. This suggests that in certain careers, for juveniles, being literate may well have been a marker in the process whereby they were singled out for better positions later in their adult careers.

Human capital's contribution to output

The two preceding sections have established that human capital was widely disseminated among the Lisbon industrial work force and that it arose essentially as the result of a process of learning that took place on the shop floor. What these sections do not tell us, though, is how important it was relative to the other factors of production. This section attempts to provide an answer to this question. It also serves to start the discussion regarding the way in which capital – skill complementarity may have been present in the industry of a retarded economy of this period.

The methodology we have adopted follows that used by Rosés (1998). The framework is that of a neo-classical production function in which firms use raw materials, labour (decomposed into raw labour and human capital) and capital. The contribution of these inputs is measured by the market value of the flow of services they provide. The aim is to estimate their respective shares in the value added by each firm in the sample, and then aggregate this with the overall industrial level. In this framework, raw labour, capital and human capital add up to value added, which is calculated in turn as the difference between gross output and the value of the raw materials required to manufacture it. It must be stressed that this does not constitute a rigorous growth-accounting exercise. It probably over-estimates the contribution of capital relative to other inputs as the latter is estimated by default, and it fails to distinguish the contribution of entrepreneurial labour from that of the rest. Likewise, it does not always distinguish fixed from circulating capital. On the other hand, its usefulness resides in providing reasonable orders of magnitude, which allow us to establish the relative importance of human capital and raw labour and their relation to capital itself.

The firm's raw labour bill is obtained by multiplying the number of workers by their average annual amount of working hours and by the hourly wage of an unskilled labourer. In the case of youths under sixteen, we have had to assume that they earned no skill premium and that their actual pay was all on account of their unskilled efforts. This involves an over-estimation because part of the subadult labour force already had some training on the job. But it is an unavoidable one given the way the wage data are presented in our source and our insufficient knowledge of the price of juvenile raw labour. Its impact on the result of this exercise should be small, however, since boy and girl workers were only 14 per cent of the total labour force and their skill premia were modest. The human capital input of each firm is derived by subtracting the amount thus obtained from the total payment to the labour force. In this way, human capital is the sum total of the skill premia actually earned by the firm's employees. The global labour bill itself was calculated by multiplying the yearly average of hours per worker by their number in each occupation and by the respective average hourly wage. Our source also supplies us with information on the value of the stock of capital. This does not permit, however, an estimation of the value of the services generated by this factor of production because we lack adequate amortization schedules and market interest rates with which to compute a realistic rental value of equipment and buildings. The contribution of total capital is obtained therefore as a residual, after deducting the value of the raw labour and human capital inputs from total value added of each firm.

The 1890 inquiry provides quantitative information on wages for only 260 firms with more than ten workers. Unfortunately, of a majority of these the record is not complete in other respects and we had to eliminate all but 39 of them in order to obtain a sample in which all the necessary indicators were present. The most frequent reason was the failure of the source to distinguish between the pay of men and women, but many other observations had to be ignored because either output or raw material values were lacking. Besides being disappointingly small, the group of firms covered in Table 2.3 cannot be claimed to be representative of the whole. The method of selection is not random and introduces some disturbing biases. For example, branches of industry such as tobacco or textiles are entirely excluded, as they were heavy users of female labour, while those that employed predominantly men, such as metal working and construction materials, are over-represented.[22]

Unlike the study by Rosés (1998), this chapter does not try to pin down the contribution of human capital to growth over a period of time. Given the lack of time series data, all we can do is estimate this factor's role in 1890 as compared with raw labour and capital. The result of the calculation is shown in Table 2.3, and at first sight may appear surprising. After all, we saw above that the vast majority of industrial occupations involved some remuneration to skill and yet we now see that the share of human capital in total value added is small (that is, lower than 20 per cent) and lower than the contribution of raw labour. Two downward influences probably account for much of this. To begin with, the composition of the sample favours activities with a relatively low input of human capital. Comparatively, human capital's share would probably rise in the case of the full sample of 260 firms. Second, several of the firms observed are, within their respective branches, among the lighter users of human capital. It is fair to presume that a more representative sample would again have given us a different picture of the balance between factors of production, and one that was less biased towards raw labour. In what follows, we must therefore bear in mind that the distribution revealed by Table 2.3 is probably a lower bound estimate of the role played by human capital in industrial production at this time.

A second lesson to draw from Table 2.3 has to do with the conspicuous position occupied by capital in the distribution of factor shares. Surprisingly,

Table 2.3 Factor shares in industrial production: Lisbon, 1890 (percentage shares of value added)

Number of firms	Number of workers	Raw labour	Human capital	Capital
39	3,081	28.2	16.0	55.8

Source: *Inquérito Industrial* (1890).

Lisbon's industry by 1890 appears to have been making considerable progress in terms of its use of technology and had gone far in replacing labour with it. It was no longer an unmechanized sector that relied mainly on the dexterity and strength of its work force.[23] If we take 'broad capital' as our reference (Rebello 1991), this substitution becomes even more striking. In this case, raw labour accounts for less than one-third of value added, while technology and skills have an overwhelming two-thirds. These results are similar to those obtained for the Catalan spinning industry in the middle of the nineteenth century (then a technologically advanced sector in the midst of what was, by international standards, already a dynamic, industrializing economy).[24] The same study reveals that the more backward, less mechanized cotton weaving industry there relied far more heavily on raw labour and traditional skills and had a capital input of less than 20 per cent. Given the interval of 40 years between these cases, it is obvious that such comparisons must be treated with caution. Similar studies are scarce and hardly ever focus on productive skills (as opposed to) as human capital, which renders it difficult to make a proper judgement about Lisbon's industry in 1890 in a comparative perspective. Nevertheless, it is still worth discussing two further questions which arise from these findings.

The first one is why should capital have played such an important role in a backward industrial setting characterized by a high cost for this input and a cheap labour force? The immediate answer may lie in the fact that Portugal was not a creator of technology but typically an importer, which obtained it from capital-abundant countries in the Western core. With little choice in the matter, Portuguese industry had to adopt the suboptimal capital-biased factor mix that Table 2.3 brings to light. The second question is why did technological progress not shift this factor mix still more towards a greater use of raw labour and away from that of human capital? This would have made particular sense for those who hold that technological progress was de-skilling and led to the replacement of human capital by raw labour and capital. Moreover, in Portugal the skilled – unskilled wage ratio was higher than in the core countries that exported industrial equipment and this should have further discouraged a higher intensity of human capital.[25] At first sight, once again the solution would be the inevitability of the country's technological external dependency, which prevented the choice of an optimal solution. In places where technology was being developed, human capital was relatively cheap. As Acemoglu (1998) has pointed out, in advanced economies human capital is more abundant and this tends to generate new technologies that favour this factor of production. It could thus simply have been this that was driving the situation described in Table 2.3, of broadly-defined capital's overwhelming presence in the production function.

An alternative way of stating the problem focuses on the internal conditions of technological choice and suggests a more plausible model. It involves considering that imported technology imposed much less of a straitjacket on how inputs were combined than the preceding remarks imply. There was in fact a greater degree of choice in the matter and this meant that the relation between capital and human capital was not simply fixed by external para-meters but depended also on the domestic circumstances that determined cost relatives and factor scarcities. This was observed by Portuguese textile managers at the time, when they lamented that although they had the most advanced machinery in the world, their weavers could not handle more than two looms each, whereas in Britain as many as three or four were operated by one person at a time (Reis 1986).

For economic historians, none of this is new. Fifty years ago, Ger-schenkron (1962 [1952]) noted that among the late developing economies of nineteenth-century Europe, where raw labour was abundant and cap-ital expensive, technological choices nevertheless favoured high capital-intensity. The reason was the persistent scarcity of suitably skilled labour which thus had to be replaced by more sophisticated equipment. This required, however, higher skills among the workers that were to mind these latest vintages of machinery, with the paradoxical result that 'the high skill premium is explained by the generally low level of skill among the growing industrial labour force' (Borodkin and Valetov 1998, p. 76).[26] The conclusion this points to is that where economic backwardness is greatest, there may be strong reasons to expect the same capital – skill complementarity which was supposed to be the hallmark of the advanced economies.

If this analysis were extended to other national cases, we might find that in backward economies capital – skill complementarity is even stronger than in the advanced ones, owing to the aforementioned 'Gerschenkron effect'. If established empirically, such a finding would be interesting for two reasons. One is that it would broaden the scope of studies in this field, which suffer from excessive narrowness in both temporal and geographic terms; the other is that it would lead to a better understanding of the underlying mechanism of this complementarity. In turn, this would help resolve the question we posed initially: is capital – skill complementarity due to certain particular technical features present only in twentieth-century industry and absent in earlier periods (Goldin and Katz 1998), or is it a more general feature of the historical process of industrialization? The Portuguese and other examples suggest that it is the latter and it is to this subject that we now turn.

Capital–skill complementarity

Although capital–skill complementarity has enjoyed a vogue in the eco-nomics literature of recent years, it has received surprisingly little attention from economic historians, except in the form of impressionistic analysis. This is particularly true of the Industrial Revolution in Britain, regarding which many have stressed the absence of this relationship. This is a tradition which harks back to Marx, although only a small part of it could be called

Marxian (Hudson 1992). Forty years ago, it was forcefully stated by Landes (1969) when he defined the technological revolution as 'the substitution of machines – rapid, regular, precise, tireless – for human skill and effort'. The lack of a detailed and critical analysis of the notion of skill explains perhaps why even now the strength of this current persists. Thus, according to Nicholas and Nicholas (1992, p. 17) 'the factory deskilled and proletarianized the work force by destroying old skills ... and [relied] on power-driven machinery which created jobs that required no formal skills or even rudimentary levels of literacy'. And the same message echoes in a recent article by Feinstein (1998, p. 651) which claims that 'skilled male craftsmen were displaced or challenged by the introduction of machinery, by change in the organisation of production ... and by employing female workers in traditional male occupations'.[27]

Not all have subscribed to this point of view, however. Capital – skill complementarity during the Industrial Revolution has long had its defenders too, and this across a broad variety of historiographic traditions. Samuel (1977), for example, has argued that 'nineteenth-century capitalism created many more skills than it destroyed, though they were different in kind from those of the all-round craftsmen, and subject to a wholly new level of exploitation'.[28] In a recent paper, Bessen (2003) has provided one of the most detailed and cogent sets of arguments in its support. His claim, based on evidence from the cotton textile industry of Lowell, Massachusetts, as far back as the 1830s and 1840s, is that the factory dispensed with certain highly paid skills but generated a demand for other ones that were different but overall no less costly to employers. The new, mechanized processes required operatives who could work faster and more steadily, given the rhythms imposed by the new sources of power. The factories also needed workers who could take decisions rapidly regarding the use of the machines they minded and, most important of all, who could avoid stoppages and breakdowns, given the high price of the new equipment relative to other factors of production.

The data at our disposal allow us to determine what this relationship was in Portugal at the end of the nineteenth century. Did the technical progress that raised capital-intensity drive a concomitant increase in the value of the package of skills that these new techniques required? By this time, a fair degree of mechanization, division of labour and investment in fixed capital had occurred in Lisbon's industrial belt but the process was uneven, as was the distribution of skills across firms. This naturally raises the question as to whether any correspondence can be found between the two variables. An OLS regression has been run in which each observation corresponds to a firm. Its result is displayed in Table 2.4. The average endowment of human capital per worker in the firm **[Hkcap]** is the dependent variable and is measured on the basis of the wage premia, as described earlier. The explanatory variable that concerns us here primarily is fixed capital per worker, also at firm

Table 2.4 Determinants of capital – skill complementarity

Dependent variable: Hkcap

(ordinary least squares regression)

Variables	
C	33,476
	(2.17)
FixKcap	0.003
	(3.090)
Hpcap	−0.591
	(−2.212)
outpcap	non-significant
Lab	non-significant
adjusted R^2	0.26
F-statistic	3.10
N	40

Note: T-statistics in brackets.
All estimations passed the White heteroscedasticity test satisfactorily.

level **[FixKcap]**, and we should consider the complementarity hypothesis validated if the sign of its coefficient were positive.

In addition, we have included several control variables, which a priori might be supposed to have had some influence on the dependent variable. Faster operation of the factory could be expected to increase skills, given the need of these in order to deal with speeded-up machinery, as pointed out by Bessen (2003). We have proxied this with a dummy for the presence of steam engines in the factory **[Steam]** or, alternatively, with the amount of steam horsepower installed per worker **[Hpcap]**. Both coefficients should be positive. An increase in throughput, meaning a greater volume of raw materials and semi-processed goods moving through the shop floor in the same unit of time, would also call for more complex equipment to handle the flow and would also step up skill requirements. It is proxied here by the value of gross yearly production per worker **[Outpcap]**, the anticipated sign again being positive. In addition, the size of the labour force **[Lab]** can be usefully included too given that the larger it is, the greater should be the division of labour within the factory.[29] This should be accompanied by an increased presence of machinery, requiring, as argued already, more skills to handle it. The coefficient of this variable should have a positive sign.

The results of the regression, in Table 2.4, are quite satisfactory, particularly bearing in mind the cross-sectional nature of the data, and they confirm the capital – skill complementarity hypothesis.[30] The main objective is met in the sense that human capital endowment **[Hkcap]** is shown to be positively and significantly related to technical progress **[FixKcap]**. The only other significant variable is steam power **[Hpcap]** but its sign is wrong (the dummy suffers from the same problem). This is probably due to collinearity and is not surprising given the association between the use of machinery and the power of steam engines installed in the factory. The labour force size **[Lab]** may suffer from the same problem as its T-statistic is too small. The throughput indicator **[Outpcap]** is also non-significant but has the right sign. It is probably not the right specification for this variable, which should be given by physical volume rather than its value. Adequate data for this could not be mustered, however.

Two final remarks are prompted by these findings. One is a note of caution regarding the shortcomings of the underlying data. The most important aspect of this is the presumable bias caused by using firms that employed male labour exclusively. The impact on these results of omitting a number of firms that were modern, highly mechanized and employed a considerable amount of female labour in conjunction with male workers unfortunately cannot be assessed. The second, more substantive comment is to underline the importance of identifying capital – skill complementarity in a period well before the moment for its onset according to the more recent literature. Moreover, in the present case, it occurred in an impoverished and late-industrializing economy which was hardly on the cutting edge of technological progress. This suggests that capital – skill complementarity is a much more enduring and widely spread feature of industrialization than has been supposed. It did not have to wait for electrification, batch production and high throughputs to make itself felt because the machines of the classic steam era generated an intense demand for new skills and abilities long before that.

Conclusion

Several conclusions emerge from this chapter which are of interest not only for the study of the role of human capital in Portuguese industrialization but in a wider context too. The first is that human capital, as measured by the premium of skill over the remuneration to raw labour, had a significant presence in Lisbon's manufacturing district at the end of the nineteenth century, despite the appearance of backwardness which the latter displayed at this time. In the second place, we have established that learning skills on the job was responsible for the most valuable part of workers' human capital. Formal educational skills contributed as well but to a small extent only. A third finding quantifies the relative importance of human capital as a production factor. It has been shown that it was much less important than fixed capital and somewhat less than raw labour. By 1890, raw labour, however, had been replaced by 'broad capital' to a very considerable extent, a fact that indicates a greater degree of technological progress than has been supposed to date. It also suggests that in late developing economies, the lack of skills may have encouraged the adoption of capital-intensive processes that in turn raised the demand for human capital. A fourth conclusion concerns the presence of capital – skill complementarity in Lisbon's industrial structure. The analysis shows that this existed even though the sector was still far from achieving the levels of technological and structural sophistication that have been considered in the literature as indispensable to this development. It also confirms that capital – skill complementarity was a thing of the First and not only of the Second or of the Third Industrial Revolutions and can be located in quite backward economies.

Notes

1 The author is grateful to Sofia Lucas Martins for valuable research assistantship, and, for comments, to Joan Rosés, Ana Rute Cardoso, Anders Nilsson, Pedro Telhado Pereira, Andrea Ichino, Jonas Ljungberg and Mar Cebrián.

2 They are referred to here as *Inquérito* 1881 and *Inquérito* 1890, respectively. The full titles are Comissão Central Directora do Inquérito Industrial (1881), *Inquérito Industrial de 1881* (Lisbon: Imprensa Nacional); and Ministério das Obras Publicas, Comércio e Indústria (1891), *Inquérito Industrial de 1890* (Lisbon: Imprensa Nacional). For a detailed survey of earlier efforts in the same direction, see Matos (1991).

3 The current scholarly consensus is that the 1881 inquiry is superior to that of 1890, a view that goes back to the 1950s (Moura 1974). In fact, the 1890 *Inquérito* enjoys several advantages over its predecessor, apart from a greater amount of quantitative information in general. It presents concrete evidence on literacy rates for all occupational categories and is much clearer in its presentation of data on the capital stock of firms. Its wage survey distinguishes within each firm by work categories and it specifies the length of the working day and year. For 1881 we are lacking all of this: for example, in wages we have only firm averages for males, females and children, respectively, regardless of occupation. There is no explanation of how they were estimated. The main comparative advantage of the first inquiry lies in its lengthy descriptions of the internal workings of many firms.

4 Our count of workers actually surveyed differs slightly from that of Mata (1999), which provides two figures: one of 14,557, on p. 132, and one of 18,476, on p. 137.

5 This relationship is also corroborated, in *Comissão Parlamentar* (1885), by an extensive survey of manufacturing wages in various provincial districts. The alternative approach of using the maximum, in lieu of the mean rate of pay, would have the advantage of revealing the upper limit of productivity of a given occupational group but would not have been representative of all its members, and therefore would have caused serious distortions in estimating global labour costs.

6 In the tin trades, the comment was 'when the worker is no longer of use, he is sacked' (p. 222); 'lathe workers receive a variable daily wage, according to their aptitudes' (p. 234); in cotton textiles, 'workers are kept on as long as they are necessary, and are dismissed without notice, when they are not' (p. 268). In the 1890 *Inquérito*, it was claimed that workers usually retired after the age of forty, when they had ceased to produce with regularity and high quality. See Vol. 4, p. 314.

7 For a similar opinion by workers representatives in the 1889 inquiry on the conditions of weavers in Oporto, see *Inquérito* (Ministério das Obras Publicas 1889), p. 27.

8   According to Lains (2003), the top ten firms of any branch in the aggregate accounted for about 15 per cent of the respective labour force.

9   The calculation based on Clark's approach would be: skill premium = (raw labour wage + 'discipline premium' + 'true' skill premium) − (raw labour wage + 'discipline premium').

10   The testimony given by trade unions in 1909 presents a harsher picture of the length of working days than the *Inquérito* of 1890 but may be biased given its origin. See Cabral (1977).

11   The real (plural, reis) was the Portuguese monetary unit. In 1890, before the onset of inconvertibility in 1891, 4,500 reis were equivalent to one pound sterling.

12   This dividing line is taken from current research on wages between 1700 and 1900 based on the archive of the royal palaces. It is equivalent to a daily wage of 400 reis, the remuneration of men engaged in heavy but unskilled work in gardens and on construction sites. See Reis (2004).

13   Based on family budgets around 1910 compiled by Martel (1911). The trade union inquiry of 1909 (Cabral 1977) presents similar cost of living levels for a family of five. For a family with a single bread-winner living decently, his hourly wage should be in the 74–120 reis range. Correction for the fact that the Lisbon cost of living in 1890 was 10 per cent lower than in 1910 has been made.

14   Owing to the methodology followed, we have been unable to class most of the workers in several important industries of this period since they were usually paid by the piece and/or included women. The most relevant of these sectors are cork, fish canning, tobacco and textiles.

15   Madureira (2001) argues that as a result of the de-skilling caused by mechanization, there was a spread of occupational designations that no longer related to the tasks performed but rather to the way in which workers were paid, e.g., 'day worker', 'operative', 'subcontractor', and so on. These categories were relatively insignificant.

16   An intermediate position is that of O'Rourke and Williamson (1997) who show that 'good schooling' (that is, either substantial school enrolment or literacy rates) can explain only a small part of European convergence during 1870–1910, although it is helpful for understanding the divergence of certain peripheral economies.

17   Madureira (2001)'s description of jobs in the textile industry at this time makes it clear that their associated skills were typically learned at work, and rarely, if ever, by means of technical education in a school. The exceptions seem to have been the schools for female spinners which were set up for a relatively short period towards the end of the eighteenth century (pp. 47–8).

18   The 1909 report on the conditions of the working class repeatedly states that the earnings of apprentices were closely related to the extent of their skills, whatever the task in which they were engaged (Cabral 1977). For earlier similar comments on the wages of boys in cotton printing, see *Inquérito* (1881), Book I, p. 78.

19   See, for example, the interview with the manager of the Xabregas tobacco factory, in Lisbon, in Vol. 1, part 2 of the 1881 *Inquérito*, p. 336.

20   Portuguese literacy statistics from the 1890 population census show that, in the Lisbon district, there was a difference of only 3 per cent between those who could read and those who could both read and write. In 1852, a survey of Lisbon industrial workers shows that these two rates were almost as close. See *Relatório* (1857).

21   Mata (1999) finds a positive relationship between literacy and technical skill, on the one hand, and the use of machinery, on the other. This conclusion is not

reached, however, by regression analysis but by visual comparison of thirteen broad industrial categories, which lump together quite a variety of jobs and of manufacturing branches.

22 The branches of industry represented are: sugar, weapons manufacturing, shoes, boiler making, carpentry, ceramics, shipbuilding, cork, leather, cotton printing, foundry work, instruments, woollens, lithography, margarine, cabinet making, glass, stone work and metal work.

23 Though still behind the industry of the advanced economies, if we take installed horsepower per worker as a proxy for technological development, it should be noted that in Lisbon, in 1890 it was 0.13, while in 1906 it was 0.45 for the whole of France and 0.79 for Britain (Dormois 1997).

24 Rosés (1998) found, over the period 1831–61, 15 per cent for human capital, 29 per cent for raw labour and 56 per cent for capital. For 1861 alone, the result was 19 per cent for human capital, 22 per cent for raw labour and 59 per cent for capital.

25 In the USA it was in the 1.7–1.8:1 range, while in Britain it was less than 2:1 (Lindert and Williamson 1980; McIvor 2001). In the Netherlands, it was even lower, according to Smits and van Zanden (1998), while in Belgium it was around 2:1 (Scholliers and Zamagni 1995). On the other hand, in Portugal it varied between 2.4 and 2.8:1 and in Russia between 2.6 and and 3.1:1 (Borodkin and Valetov 1998). For a model that helps to explain why these differences among countries endured in the long run, see Arora (2001).

26 Chary and Hopenhayn (1991) have shown that recent technologies raise the wage profile of the workers that use them relative to the older technologies. This would make the described human capital effect even more pronounced.

27 For similar views regarding the Portuguese situation, see Mónica (1979).

28 See also, as examples, Mathias (1969) and Tranter (1981).

29 Following Berg (1994), a negative sign, on the other hand, would indicate the presence of a 'flexible specialization' type of organization, as described by Sabel and Zeitlin (1985), where skill intensity was high, productive units were small and fixed capital was low.

30 As Schön (2000) has shown, the relationship between human capital and technology in a given firm or sector is apt to vary over the medium-term economic cycle, depending on whether they are undergoing a period of 'innovation' or one of 'rationalization'. Our data, which only provide a snapshot in 1890, do not allow us to take this into consideration, and this should weaken the capital – skill link that we are trying to capture. The same is apt to happen because the sample covers a variety of technologies, some more, others less capital-intensive, and therefore with varying degrees of capital – skill complementarity.

References

Abramovitz, M. (1993), 'The Search for the Sources of Growth: Areas of Ignorance, Old and New', *Journal of Economic History*, 53, pp. 217–43.

Acemoglu, D. (1998), 'Why Do New Technologies Complement Skills? Directed Technical Change and Wage Inequality', *Quarterly Journal of Economics*, 113, pp. 1055–89.

Arora, S. (2001), 'Health, Human Productivity and Long-Term Economic Growth', *Journal of Economic History*, 61, pp. 699–749.

Berg, M. (1994), *The Age of Manufactures: Industry, Innovation and Work in Britain* (Oxford: Oxford University Press).

Bessen, J. (2003), 'Technology and Learning by Factory Workers: The Stretch-Out at Lowell, 1842', *Journal of Economic History*, 63, pp. 33–64.

Boletim do Trabalho Industrial (1906–1917).

Boot, H. M. (1995), 'How Skilled were Lancashire Cotton Factory Workers in 1833?' *Economic History Review*, 48, pp. 283–303.

Borodkin, L. and T. Valetov (1998), 'Modelling Wage Inequality in Russian Industries: 1880–1914', in Leonid Borodkin and Peter H. Lindert (eds), *Trends in Income Inequality during Industrialization* (Madrid: Fundación Fomento de la Historia Económica).

Cabral, M. V. (ed.) (1977), *O Operariado nas Vésperas da República (1909–1910)* (Lisbon: GIS).

Carqueja, B. (1916), *O Povo Português* (Porto: Livraria Chardron).

Chary, V. V. and H. Hopenhayn (1991), 'Vintage Human Capital, Growth and the Diffusion of New Technology', *Journal of Political Economy*, 99, pp. 1, 142–65.

Clark, G. (1994), 'Factory Discipline', *Journal of Economic History*, 54, pp. 128–63.

Comissão Central Directora do Inquérito Industrial (1881), *Inquérito Industrial de 1881*, (Lisbon: Imprensa Nacional).

Comissão Parlamentar para o Estudo da Emigração Portuguesa (1885), *Documentos Apresentados à Camara dos Senhores Deputados e por ela Mandados Publicar 1886* (Lisbon: Imprensa Nacional).

Cordeiro, J. M. L. (1996), 'Empresas e Empresários Portuenses na Segunda Metade do Século XIX', *Análise Social*, 136–7, pp. 313–42.

Crafts, N. (1995), 'Exogenous or Endogenous Growth? The Industrial Revolution', *Journal of Economic History*, 55, pp. 745–72.

Dormois, J. P. (1997), *L'Économie Française Face à la Concurrence Britannique à la Veille de 1914* (Paris: L'Harmattan).

Feinstein, C. (1998), 'Pessimism Perpetuated: Real Wages and the Standard of Living in Britain during and after the Industrial Revolution', *Journal of Economic History*, 58, pp. 625–58.

Filer, R. K., D. S. Hamermesh and A. E. Rees (1996), *The Economics of Work and Pay* (New York: HarperCollins).

Gerschenkron, A. (1962), *Economic Backwardness in Historical Perspective* (Cambridge, MA: Harvard University Press).

Goldin, C. (2001), 'The Human-Capital Century and American Leadership: Virtues of the Past', *Journal of Economic History*, 61, pp. 263–92.

Goldin, C. and L. Katz (1998), 'The Origins of Technology – Skill Complementarity', *Quarterly Journal of Economics*, 113, pp. 693–732.

Goldin, C. and L. Katz (2000), 'Education and Income in the Early Twentieth Century: Evidence from the Prairies', *Journal of Economic History*, 60, pp. 782–818.

Huberman, M. (1986), 'Invisible Handshakes in Lancashire: Cotton Spinning in the First Half of the Nineteenth Century', *Economic History Review*, XLVI, pp. 987–98.

Huberman, M. (1991), 'How Did Labour Markets Work in Lancashire? More Evidence on Prices and Quantities in Cotton Spinning, 1822–1852', *Explorations in Economic History*, 28, pp. 87–120.

Hudson, P. (1992), *The Industrial Revolution* (London: Arnold).

Lains, P. (2003), *Os Progresso do Atraso. Uma Nova História Económica de Portugal* (Lisbon: Imprensa de Ciências Sociais).

Landes, D. (1969), *The Unbound Prometheus: Technological Change and Industrial Development in Western Europe from 1750 to the Present* (Cambridge: Cambridge University Press).

Lindert, P. and J. G. Williamson (1980), *American Inequality: A Macroeconomic History* (New York: Academic Press).

Maddison, A. (1995), *Monitoring the World Economy, 1820–1992* (Paris: OECD).

Madureira, N. L. (ed.) (2001), *História do Trabalho e das Ocupações. Vol. I: A Indústria Textil* (Lisbon: Celta).

Martel, S. (1911), 'A Alimentação das Classes Pobres e suas Relações com o Trabalho', *Boletim do Trabalho Industrial*, 44, pp. 3–42.

Mata, M. E. (1999), 'Indústria e Emprego em Lisboa na Segunda Metade do Século XIX', *Ler História*, 37, pp. 127–46.

Mathias, P. (1969), *The First Industrial Nation: An Economic History of Britain, 1700–1914* (London: Methuen).

Matos, A. M. C. (1991), 'A Indústria no Distrito de Évora, 1836–90', *Análise Social*, 112–13, pp. 561–81.

Matthews, R. C. O., C. H. Feinstein and J. C. Odling-Smee (1982), *British Economic Growth 1856–1973* (Stanford, CA: Stanford University Press).

McIvor, A. J. (2001), *A History of Work in Britain, 1880–1950* (Basingstoke: Palgrave).

Mincer, J. (1974), *Schooling, Experience and Earnings* (New York: Columbia University Press).

Ministério das Obras Publicas, Comércio e Indústria 1891, *Inquérito Industrial de 1890* (Lisbon: Imprensa Nacional).

Ministério das Obras Publicas, Comércio e Indústria 1889, *Inquérito sobre o Estado da Tecelagem na Cidade do Porto e Situação dos Respectivos Operários* (Lisbon: Imprensa Nacional).

Mitch, D. (1992), *The Rise of Popular Literacy in Victorian England: The Influence of Private Choice and Public Policy* (Philadelphia: University of Philadelphia).

Mitch, D. (1999), 'The Role of Education and Skill in the British Industrial Revolution', in Joel Mokyr (ed.), *The British Industrial Revolution: An Economic Perspective* (Boulder, CO: Westview Press).

Mónica, M. F. (1979), 'Uma Aristocracia Operária: Os Chapeleiros (1870–1914)', *Análise Social*, 60, pp. 859–945.

Mónica, M. F. (1987), 'Capitalistas e Industriais (1870-1914), *Análise Social*, XXIII, pp. 795–863.

Moura, F. P. (1974), *Por Onde Vai a Economia Portuguesa?* (Lisbon: Seara Nova).

Nicholas, S. J. and J. M. Nicholas (1992), 'Male Literacy, "Deskilling" and the Industrial Revolution', *Journal of Interdisciplinary History*, XXIII, pp. 1–18.

Nunes, A. B. (1989), 'População Activa e Actividade Económica em Portugal dos Finais do Século XIX à Actualidade', PhD dissertation, Technical University of Lisbon.

O'Rourke, K. and J. G. Williamson (1997), 'Around the European Periphery 1870–1913: Globalization, Schooling and Growth', *European Review of Economic History*, 1, pp. 153–91.

Pedreira, J. M. (1990), 'Social Structure and the Persistence of Rural Domestic Industry in XIXth Century Portugal', *Journal of European Economic History*, 19, pp. 521–48.

Pereira, M. H. (2001), *Diversidade e Assimetrias: Portugal nos Séculos XIX e XX* (Lisbon: Imprensa de Ciências Sociais).

Poinsard, L. (1910), 'Le Portugal Inconnu. II. L'Industrie, le Commerce et la Vie Publique', *La Science Sociale*, 25, pp. 231–437.

Rebelo, S. (1991), 'Long-run Policy Analysis and Long-run Growth', *Journal of Political Economy*, 99, pp. 500–21.

Reis, J. (1986), 'The Industrialization of a Late and Slow Developer: Portugal, 1870–1913', *Rivista di Storia Económica*, 3 (International Issue), pp. 67–90.

Reis, J. (forthcoming), (2004), 'O Trabalho no Século XIX', in P. Lains and J. A. F. Silva, *História Económica de Portugal* (Lisbon: Imprensa de Ciências Sociais).

Relatório da Repartição de Manufacturas do Ministério de Obras Públicas, Comércio e Indústria Apresentado à Câmara dos Srs. Deputados pelo Ministro e Secretário de Estado Respectivo (1857) (Lisbon: Imprensa Nacional).

Rodrigues, M. F. and J. M. A. Mendes (1999), *História da Indústria Portuguesa da Idade Média aos nossos Dias* (Mem Martins: Europa-América).

Rosés, J. R. (1998), 'Measuring the Contribution of Human Capital to the Development of the Catalan Factory System (1830–1861)', *European Review of Economic History*, 2, pp. 25–48.

Sabel, C. and J. Zeitlin (1985), 'Historical Alternatives to Mass Production: Politics, Markets and Technology in Nineteenth Century Industrialization', *Past and Present*, 108, pp. 133–76.

Samuel, R. (1977), 'Workshop of the World: Steam Power and Hand Technology in Mid-Victorian Britain', *History Workshop*, 3, pp. 6–72.

Scholliers, P. and Zamagni, V. (1995), *Labour's Reward. Real Wages and Economic Change in Nineteenth and Twentieth-Century Europe* (Aldershot: Edward Elgar).

Schön, L. (2000), 'Electricity, Technological Change and Productivity in Swedish Industry, 1890–1990', *European Review of Economic History*, 4, pp. 175–94.

Smits, J. P. and J. L. van Zanden (1998), 'Industrialiazation and Income Inequality in the Netherlands, 1800–1914', in L. Borodkin and P. H. Lindert (eds), *Trends in Income Inequality during Industrialization* (Madrid: Fundación Fomento de la Historia Económica).

Tranter, N. L. (1981), 'The Labour Supply 1780–1860', in R.Floud and D. N. McCloskey (eds), *The Economic History of Great Britain since 1700* (Cambridge: Cambridge University Press), pp. 204–26.

3

Engineering Expertise and the Canadian Exploitation of the Technology of the Second Industrial Revolution

Marvin McInnis

Canada was arguably the most successful exploiter of the new technology of the Second Industrial Revolution. The concatenation of more scientifically-based technological developments occurring late in the nineteenth century gave a great boost to economic performance throughout the European world. After a couple of decades of languishing economic growth, the pace of change was invigorated right at the end of the nineteenth century so that many countries entered the new century in a vibrant condition, developing more rapidly as the new technology boosted productivity in many areas of their economies. In the language of the New Growth Theory, here was a technological shock of great consequence. Mass production of cheap steel led the way, supplanting the older iron technology. Then there was the electrical revolution, providing not only a new form of prime mover to power manufacturing but an array of technically new, electrically-based processes, and an assortment of new consumer products as well. Scientific chemistry also came into play. Commodities came to be produced by chemical synthesis, and entirely new elemental combinations were found to have valuable uses. Finally came the internal combustion engine. It found application in the early twentieth century not only in powering automobiles but in agricultural machinery, marine uses, and stationary power sources as well. In short, there was a great burst of technological innovation that has long been recognized to have invigorated the turn-of-the-century economy.

These remarkable technological developments greatly enhanced productivity. That was not their only consequence since they also added greatly to the range of products that people could enjoy, and they allowed people to do things such as fly and talk to those on other continents. The revolutionary nature of these developments makes productivity comparisons exceedingly difficult. Nevertheless, one of the things they did was to make possible the production of goods using fewer or cheaper resource inputs. In the standard language of economists they brought favourable shifts in production functions. That is why they are commonly looked upon as an important source of economic growth. Caution must nevertheless be exercised in evaluating the changes since in so many cases there were such important modifications in the very nature of the products. With what does one compare the instantly switched-on, high lumen electric light, or the unit-drive electrical motor? Is the automobile just a faster, more comfortable carriage? Some sense of the dimension of change can be gained for reasonably comparable products. Open hearth steel, for example, could be produced with fewer resource inputs than crucible steel. Paper from wood pulp was much cheaper than paper from rags. An outstanding example was aluminium produced by the Hall-Héroult process, but that in turn required vast amounts of cheap electricity. These types of development, though, exemplify what we mean by technologically-based increases in total factor productivity. The surge of technological developments that occurred in the last quarter of the nineteenth century and early years of the twentieth (often referred to as the Second Industrial Revolution) was especially fruitful as regards productivity and growth enhancing developments.

This chapter builds upon the premise that Canada benefited greatly from these technological developments, more so arguably than any other economy. The question of primary interest is how it was that Canada was so well able to exploit this new technology and to benefit so much from it. The absorption and implementation of new technology is not a simple or automatic matter. It does not happen without cost. Resources have to be used to incorporate the new technology into the production structures of any economy. New Growth Theory makes this essentially a matter of investing in human capital, but that is surely too vague and simplistic a way of putting it. In reality, the process of absorption and exploitation of new technology is almost certainly a lot more complicated. It may have many dimensions. Still, we need to strive to understand it.

Consider first the case that Canada was in the forefront of exploitation of the technology of the Second Industrial Revolution and able to accomplish more with it than other economies of the European world. It is not necessary to prove that in order to validate this chapter. It really only matters that the new technology played an important role in the growth of the Canadian economy in the years between 1897 and the outbreak of the First World War.[1] That in itself would make an investigation worth while. It appears, however, that the Canadian case was outstanding, and that Canada may have accomplished more with this technology than other national economies. If so, that gives special force to an exploration of the Canadian case.

The electrical technology appears to have been especially beneficial. It freed the Canadian economy from a long-standing limitation of energy resources in the era of steam, since coal was found only at the eastern and western extremities of the country, and not at all in the most heavily-populated central heartland.[2] Hydraulic resources, with which electricity could be generated, abounded in Canada. Chemical processes allowed Canadians to make use of other abundant natural resources.[3] The new technology generated a range of new products for Canadians to manufacture, and new ways to make them. The main impact of the internal combustion engine and the automobile industry to which it gave rise came almost entirely after the period under consideration here, but steel was another matter. The new technologies of producing steel by either the Bessemer converter or the open hearth process had been invented some time before 1897, but widespread implementation outside Britain had been limited before the late years of the nineteenth century. Modern steel making was being introduced in the USA by the 1880s so, by 1897, Canada was something of a late-comer to the steel revolution. Nevertheless, the big expansion of the Canadian steel industry at the turn of the century played a major role in the sharp acceleration of the country's economic growth, and is squarely a part of Canada's participation in the Second Industrial Revolution.

The economies of European countries had progressed only slowly over the years from the cyclical downturn of the mid-1870s through to the mid-1890s. The subsequent couple of decades was generally a period of rapid industrialization and economic growth. In that setting, the rate of growth of real per capita income in Canada over the whole of the period from 1897 to 1913 exceeded that of any other country. As elaborated more fully elsewhere (McInnis 1999), the period of Canada's most notable reliance upon, and greatest benefit from, the technology of the Second Industrial Revolution was the decade 1897 to 1907. There was a short but fairly sharp recession in 1907 that marks the end of a phase of growth in the Canadian economy. In the years that followed the economy continued to grow rapidly in absolute terms, but in a way that was much more related to agricultural settlement on the western plains. At the same time the rate of per capita income growth in this later period was substantially lower than in the 1897–1907 decade, and no different from the rates of growth achieved by most industrial economies at that time.

It is the 1897–1907 decade that stands out as Canada's glittering era of economic development. Average annual rates of growth of real per capita income for the two periods 1897–1907 and 1897–1913 are shown in Table 3.1 for the set of advanced economies for which such data are available. The salient point is that Canada leads the pack whichever period is considered. It is in the 1897–1907 decade, however, that Canada's record looks so outstanding. It grew significantly faster than Australia, New Zealand or the USA, and much more rapidly than the rest of the European economies. Its closest comparator was Italy, which grew at an average rate over that period of 3.85 per cent per annum while the growth rate of Canadian real per capita income was 4.43 per cent. The USA grew at just 3.0 per cent.[4] Between 1870, when reliable national income data become available, and 1913, Canadian real per capita income converged upon the level of the USA. Two-thirds of the narrowing of the gap came in the years between 1897 and 1907.

Especially since I am arguing that Canada's success derived mainly from industrialization, it would be instructive to compare rates of industrial

Table 3.1   Comparative real per capita income growth, selected industrialized countries (average annual rates of growth)

	1897–1907	1897–1913
Canada	4.43	3.63
Australia	3.05	2.58
Belgium	0.96	1.02
Denmark	2.01	1.97
Finland	1.23	1.65
France	1.52	1.75
Germany	1.62	1.74
Italy	3.85	3.22
Japan	2.24	1.68
Netherlands	−0.44	0.37
New Zealand	3.05	1.67
Norway	0.91	1.66
Sweden	2.28	2.29
UK	0.93	0.90
USA	3.00	2.15

Sources: Sweden, unpublished revised estimates from Lennart Schön; all other countries, Maddison (1995).

growth in the post-1897 period. This has to be done with considerable caution since there is not presently available a conventional index of industrial production for Canada that covers the years under consideration. Rates of growth of industrial production can be calculated for seven leading countries, including the USA and Japan as well as the prominent European nations.[5] The Canadian measure that is available, the growth of the industrial component of gross national product (GNP), is not precisely the same thing but can be used for a rough comparison.[6] Over the decade 1897 to 1907, Canadian industrial growth (at 8.02 per cent per annum) was much faster than that of any of the other countries for which comparisons can readily be made. The demographic base of the Canadian economy was also growing faster than that of other economies and that should be taken into account. In Table 3.2, Canadian per capita industrial growth is compared with that of seven other countries for which data can readily be obtained. For Canada, the extent to which the growth of real industrial product exceeded the growth of population stands out, well above that of any European country and almost a full percentage point above that of the USA, the country commonly thought to be the world leader in industrialization at this time. It has been more common for writers to look upon the entire period from 1897 to 1913 as a piece and so in Table 3.2 that comparison is given as well. It is evident that the rate of growth of Canada's industrial production tapered off after 1907, while its population growth rate went up. As a consequence,

Table 3.2 International comparative growth of industrial output (average annual rates of growth)

Country	1897–1907			1897–1913		
	Industrial production (A)	Popu-lation	(A) per capita	Industrial production (A)	Popu-lation	(A) per capita
Canada	8.02	2.27	5.75	6.29	2.52	3.77
USA	6.68	1.88	4.80	5.76	1.88	3.88
Austria	4.00	1.03	2.97	3.33	0.99	2.36
Germany	4.07	1.47	2.60	4.06	1.41	2.65
France	1.75	0.15	1.60	2.57	0.17	2.40
Italy	5.45	0.74	4.71	4.06	0.74	3.32
Sweden	4.75	0.72	4.03	3.45	0.75	2.70
Japan	3.11	1.13	1.98	4.05	1.21	2.84

Sources: Canada, calculated from Urquhart (1993); all other countries from Mitchell (1998).

looking at the entire 1897–1913 period, the per capita growth of industrial production in Canada was not greater than in the USA. Canadian industrialization was still more rapid than that experienced by European nations; more rapid than that of Germany, the nation that is often looked upon as the epitome of industrial development at that time; more rapid than that of Sweden, a country of similar size (in population terms) and one that was similarly utilizing the new technology to good effect. The closest to Canada's achievement was that of Italy.

Canadian economic historians have traditionally emphasized the settlement of the wheat growing region of the Canadian west as the driving force behind Canada's remarkably high rate of economic growth in the early twentieth century. Indeed, the period is usually referred to as the 'wheat boom' (see, for example, Fowke 1957; Easterbrook and Aitken 1965, especially ch. 20; Urquhart 1986). I argue that Canadian economic growth, 1897–1907, was led by industrialization and especially by an industrialization that focused on those industries most reflective of the new technology of the Second Industrial Revolution.[7] Over the period in question the net output of manufacturing industry as a whole, in real terms (see note 6), grew at an average rate of 8.0 per cent per annum. This remarkably high rate of growth was admittedly a passing phase in the country's development, and was not something that could be long sustained. It was a phase that featured the high, early expansion rates of several industries that were especially affected by the new technology. The point of my argument is that at some time or another most economies experience a phase of rapid growth.[8] What calls for emphasis in the Canadian case is that the rate of growth achieved during this spurt was extraordinarily high. Furthermore, it was not just the infant growth spurt of a newly industrializing nation. Already by 1890 Canada was substantially industrialized, a point insufficiently recognized either by Canadians or by scholars from other countries.[9] In that year only three other countries in the world had higher per capita industrial output than Canada. They were the UK, the USA, and Belgium. By that standard Canada was, by 1897, well along the route to being an advanced industrial nation. Its remarkably high rate of industrial growth cannot be passed off as simply the early stirring of a newcomer to the industrial world.

The structural pattern of this industrial growth is shown in Table 3.3. There I contrast the growth of the industries most clearly affected by the new technology with a set of industries that were much less influenced. The first group should be self-explanatory, but the second group comprises industries little affected by the new technology. This is not to say that there was no influence, but that it was not a predominant feature of the industries in question.[10] The industries listed in the upper panel of the table grew much more rapidly than those in the lower panel over the period under examination. In addition to the growth rate the table also shows the weight attached to each industry: the percentage contribution it made to GNP originating in manufacturing in 1907. Attention to these weights is important in considering the contribution made by the growth of these industries in the overall growth of the economy. The net output of the large iron and steel industry grew at a remarkable 16.7 per cent per year over the decade. The much smaller electrical apparatus industry grew at an even higher rate (19.3 per cent), the chemical industry at 10.3 per cent, and the non-ferrous metal smelting and refining industry at 15.8 per cent. These are stunning rates of growth.[11]

Table 3.3 Growth rates and size of industrial sectors in Canada, 1897–1907

Technologically impacted industries	Average annual growth rate	Manufacturing (%) output
Rubber products	9.64	0.81
Paper	8.25	2.44
Iron and steel	16.70	21.71
Non-ferrous metals	15.80	5.56
Electrical apparatus	19.30	1.44
Chemicals	10.30	2.77
Industries little affected		
Primary textiles	1.68	3.37
Clothing	6.17	9.61
Leather products	−0.84	4.93
Wood and wood products	7.10	16.78
Printing and publishing	3.00	3.05

Source: Calculated from Urquhart (1993).

One might wonder whether in the case of an industry such as non-ferrous metals (in Canada copper, lead, nickel and zinc) attention to value added in the industry diminishes the apparent importance of the industry since the principal inputs were closely associated and all part of Canadian GNP. The gross output of that industry was a little more than double its contribution to GNP. The rate of growth of gross output, however, was only slightly higher than the rate of growth of value added (16.7 per cent compared with 15.8 per cent). The rates of growth of these new-tech industries were spectacular, and much higher than those attained by older and more traditional manufacturing sectors. Wood and wood products manufacturing, for example, which had previously been the single largest manufacturing sector, grew at a healthy, but nevertheless lower, rate of 7.1 per cent. Printing and publishing, in an era that saw the great expansion of the daily newspaper, grew by only 3 per cent. Some of the new industries, such as chemicals and electrical apparatus, were quite small in size and did not carry much weight in the aggregate national picture. The steel industry, however, was large and carried a lot of weight. That is why it is given special attention in what follows.

One further point relating to the strong and successful performance of the Canadian economy has been made in a recent article by Ian Keay (2000). He shows that from 1907 onwards, in a representative selection of manufacturing industries, total factor productivity in Canada was not significantly different from in the USA. That is in contrast to what was long believed to be a gap between Canadian and US performance, a conclusion based on comparisons of output per worker. Lower labour productivity in Canada was evidently offset by higher productivity of capital and, in some cases, of material inputs; and also by technological adaptation on the part of Canadian industries to differences in factor prices. Canadian real per capita GDP fell below that in the USA partly because labour productivity was lower in Canada, but also presumably because of lower productivity in industries other than manufacturing.[12]

If, as I argue, Canadian economic development in this turn-of-the-century period was founded very largely upon the technology of the Second Industrial Revolution, a pressing question is how did the Canadians do it? How did they gain the requisite knowledge to carry off this impressive accomplishment?

Entrepreneurs and engineers

Two sorts of people were involved in the successful transfer of technology to Canada:[13] these were entrepreneurs and engineers. The country had to have an adequate supply of both. There had to be entrepreneurs who were knowledgeable about the potentialities of the new technology and who had the vision to initiate projects employing it. One can ask how those people became aware of the new prospects and how they knew enough about them to initiate ventures that exploited this technology. This presupposes the existence of a cadre of educated, attentive, and well-informed entrepreneurs. Before turning to the question of the supply of engineers, let us briefly consider first the matter of entrepreneurs. Without focusing specifically on the period under consideration here, it is often generally supposed that the Canadian economy has been weakly supplied with effective entrepreneurs.

Canada may or may not have lacked sufficient entrepreneurs to have initiated many of the projects that exploited the new technology. The issue has never been adequately addressed from this perspective. There is some anecdotal evidence. We can cite instances but we lack quantitative measures or frequencies. What can be said is that whether or not Canada was spawning the needed entrepreneurs of its own, there was an abundance of Americans prepared to do the job. We see good examples of this in the steel industry, an industry which played such a large role in the development of the period. In the case of steel, entrepreneurs from the USA were involved in all the major ventures that provided Canada with an integrated steel industry. It was a group of American businessmen who determined that by the early 1890s it had probably become profitable to produce steel in a modern integrated plant located at Hamilton, Ontario. Iron ore in America was increasingly coming from further west, via the lake shipping of the Great Lakes. The enlargement of the Welland Canal in the late 1880s meant that coal or coke from Ohio and Pennsylvania could also be brought in cheaply. At about the same time that it became attractive to produce steel at Buffalo and Cleveland and Gary, Hamilton became a feasible location. A substantial market for steel had emerged in Canada. Steel was being rolled at Hamilton and at Montreal, and moreover it was a tariff-sheltered market. In addition, the government of Canada offered the additional inducement of a bounty on the production of pig iron. In 1893 American enterprise, boosted by a healthy local subsidy, initiated the erection of a steel plant at Hamilton. The severe depression of the mid-1890s, coupled with natural disaster (the newly erected blast furnace was blown down in a wind) led to the abandonment of this steel venture. At that point the project was taken up by local entrepreneurs and eventually brought to fruition. The driving force then was William Southam, the local newspaper publisher, financially backed by George Gooderham, of the Toronto distilling family. They brought into existence what would eventually be known as STELCO. The initial impetus, however, had come from knowledgeable American entrepreneurs.[14]

At Sault St. Marie, Ontario, it was an American visionary, Francis Clergue, who attempted to build a great industrial empire involving hydroelectricity, paper and chemicals, as well as steel.[15] In his mind Clergue was prepared to implement the entire Second Industrial Revolution at this remote location. Mainly what the Sault had to offer was an abundance of pulpwood and a great hydroelectric power site. There was also the possibility of a large supply of iron ore.[16] Clergue had financial backing from metropolitan America, and hefty subsidization from the government of Canada. That, in the end, the project did not work out very successfully is another matter;[17] for a while it added greatly to Canada's industrial might.

Another American, Henry Whitney of Boston, had gained control of a major part of Nova Scotia's coal resource. He was blocked in his endeavour to fuel Boston with Nova Scotia coal by the Bostonians' recognition that Nova Scotia coal was dirty and sulphurous. Anti-pollution regulations were enacted to bar its use in the city. Whitney had to look for other uses for his coal and seized upon the idea of steel. Steel was the exciting new technology of the day and its prospects in Nova Scotia were enhanced by the recent discovery of iron ore at tidewater in nearby Newfoundland. In Whitney's scheme there was the prospect of building a great, export-oriented steel enterprise. That he did not know about the problems of quality and extraction cost of Nova Scotia coal, and was not able to keep investment expenditures under sufficient control to make Dominion Steel and Coal Corporation (DOSCO) a really profitable enterprise, is another matter. Whitney quickly sold out to a consortium of Toronto investors. They were prepared to put up funds for what they thought was a promising opportunity but they had not conjured up the idea in the first place. From the perspective of this chapter the important point is that again in this case the entrepreneurial initiative had come originally from the USA.

There is a counterbalance to the foregoing story. The real pioneer in modern steel making in Canada was Nova Scotia Steel and its corporate forebears. In 1882 local entrepreneurs opened Canada's first modern steel furnace at Trenton, Nova Scotia, well before the giant enterprises came into existence. This was a product of local, Canadian entrepreneurship, the Drummond and McGregor families of Pictou, with the financial backing of John Stairs of Halifax. It was a successful and profitable firm. It utilized the new open hearth technology. Traditionally, however, the achievement of Nova Scotia Steel has been minimized because it was not an integrated plant. Like most of Britain's 'modern' steel industry, it produced open hearth steel from pig iron smelted elsewhere. It nevertheless represented a successful pioneer endeavour in the transfer of technology, promoted by local entrepreneurs and carried through primarily by local expertise.

Other industries that were built upon important elements of the new technology may provide additional examples of both imported and indigenous entrepreneurship.[18] What we gain from them are stories, often interesting stories, but not systematic evidence that can be used to support a general explanation. The entrepreneurial role in bringing the Second Industrial Revolution to Canada remains, for the present, still largely unexplored. There remains a second issue. How did Canada obtain the technical expertise to carry through the implementation of the new technology? This is a question that concerns the supply of highly skilled manpower: scientists, engineers, and similarly highly skilled persons. The most readily identifiable are the engineers. They may not constitute the whole story but, through their expertise, they comprise a large part of the story. Where and how did Canada obtain the cadre of engineers needed to exploit the new technology of the Second Industrial Revolution? The case may be parallel to that of the entrepreneurs; if engineers were not being produced in Canada, they could always be brought in from elsewhere, particularly the USA. It would be interesting to know if that is really the way it happened.

Economists and economic historians have for the most part paid little attention to the role of engineering expertise in the development of economies. That is especially true for Canada. It has long been customary to emphasize the central role of technological progress in the growth of economies, and economic historians are wont to make the point that they were telling that story long before Solow and other economists came to emphasize it. Initial inventors are identified and accounts given of pioneer applications of the technology but little is said about how the technology diffused or about how the many imitators were able to build entire industries. There is a literature on the emergence of engineering as a profession, and on engineering education. There is also a literature on the history of science that makes frequent reference to engineers. Few writers, however, have directly addressed the issue of the contribution of engineers to economic development. An exception is Ahlström (1982), whose slender book on higher technical education and the engineering profession in the late nineteenth and early twentieth centuries in England, France, Germany and especially in Sweden stands out as a pioneering contribution. It offers some international comparisons against which the Canadian case can be examined. Rosenberg (1998) has provided us with a careful look at the development of chemical engineering. Besides the early work of Blank and Stigler (1957), which covers engineering along with other scientific personnel, Edelstein (2001) has recently written about the supply of engineers in New York State coming from that state's institutions of higher learning. His investigations of the US case are still at a relatively early stage, but the same author had previously written about the supply of engineers in Australia (Edelstein 1988). On the whole, however, there has been rather little attention in the literature either to the supply of engineering talent or to its role in the economy. The studies mentioned put their emphasis on engineers produced by institutions of higher learning. Those institutions kept records that are at least in principle accessible, and they often published reports. At the turn of the twentieth century, however, it may be that at least half of the practising engineers in North America were not the products of formal education in engineering at the university level (Mann 1918).

With an interest in numbers of people pursuing particular occupations we automatically turn to the decennial census of Canada. The occupational categories reported in the Canadian censuses of 1901 and 1911 identify several types of engineer. In 1901 they totalled a mere 2,076, and that included surveyors who were grouped with civil engineers and would almost certainly have outnumbered the engineers in that category. Engineering was not a common occupation and the number reported would have amounted to a mere 0.39 engineers per thousand people in the nation. The census also identified a few other scientifically oriented people designated as inventors, chemists (but compounded to an unknown extent with pharmacists), and a relatively large number (2,583) of metallurgists and assayers. By 1911 the number of engineers of all types had increased to 5,610, in per capita terms almost a doubling (to 0.78 per thousand population), and that figure no longer grouped surveyors (who numbered 1,729) with the civil engineers. This clearly was a rapidly expanding occupation. Of the engineers enumerated in the 1911 census, 44 per cent were not Canadian-born. It would appear that Canada was heavily dependent upon immigrant engineers although that dependence is to some degree over-stated since some of the foreign-born would have come to Canada as children and have been trained or educated as engineers in this country. In 1911, some 37 per cent of the non-agricultural male work force was foreign-born, so engineers were not much out of line with workers of all sorts in their inclusion of people born out of the country. Although I have argued elsewhere (McInnis 1994) that, in the four decades leading up to 1900, Canada was essentially a nation of emigration, not largely of immigration, there had been a reversal in the years immediately before 1900, and between that year and 1911 immigrants had arrived in large numbers. They included engineers along with farmers, craftsmen and labourers. The older generation of engineers practising in 1911 also included a good number who had come to this country in the heavy immigration in the years before 1860.

It would be attractive to compare Canada to the USA with regard to the relative numbers of engineers in these early years; however, incomparabilities in census classifications make this problematic. In the US censuses of 1890, 1900 and 1910 surveyors are grouped with civil engineers. Electrical engineers are grouped with the much larger number of electricians. Blank and Stigler (1957) made adjustments to the census numbers to attempt to bring them closer to comparability over time. They deducted an estimated number of surveyors from the count of civil engineers and put forward figures for the total number of engineers in the USA in 1900 and 1910.[19] Those can be compared with the Canadian census numbers for 1901 and 1911. There is a further question of what base should be used to normalize the number of engineers for the purpose of international comparison. Simply to look at numbers of engineers per capita would fail to take into account the greater preponderance of agriculture in the Canadian economy and the lesser need for engineering services in an agricultural economy. It was also the case that the Canadian birth rate was higher and the proportion of young, non-working people higher than in the USA. A more appropriate base for comparative purposes might be the non-agricultural work force of each country. In 1900 the USA had 3.0 engineers per thousand male non-farm workers; in 1910 it had 4.6. In Canada, the ratios for 1901 and 1911, respectively, were 2.48 and 3.88. This comparison does not suggest any great lag of Canada behind the USA in the density of engineers. Furthermore, Canada was narrowing the gap. The evidence presented by Ahlström (1982) indicates that the density of engineers in France and Germany was considerably greater than in the USA or Canada.

More than just summary census data are needed to inform us about how Canada was supplied with engineering talent in these early years. In what follows I explore three sources of information: the development of engineering schools and the numbers of their graduates; the records of persons notable for their engineering accomplishments, to be found in the *Dictionary of Canadian Biography*; and a sample of engineers practising in Canada in 1911.

University education in engineering in Canada

In the mid-nineteenth century engineers trained as apprentices, much like any other craft. Civil engineering, in its original sense of a contrast with military engineering, was as much as anything an extension of surveying. Some of Canada's most successful engineers were self-taught. Outstanding examples of that are Benjamin Chaffey and George Chaffey Jr, who are the principal objects of another investigation that I have underway.[20] Both did outstanding work as engineers, yet had no formal training or even much in the way of apprenticeships. Canada also drew notably on immigrants trained in Europe. Well known examples here are Sandford Fleming and Casimir Gzowski.[21] It was the construction of canals and railways that brought forth the greatest demand for persons with engineering capability. Science, in general, was only beginning to make its way into university education. As technology became more scientifically informed, university education in science came to be more important for scientists and engineers. University programmes to provide for that education began to emerge.[22]

McGill University at Montreal pioneered a diploma course in applied science in 1857 but in 1863 it was discontinued when not a single student enrolled.[23] The programme was revived in 1871 and from that time forward McGill continuously offered instruction in engineering. By 1874 the McGill programme had 33 students, and four years later more than double that. Five students graduated in engineering from McGill in 1874. The early 1870s saw a flurry of interest in applied science education across Canada.[24] In 1872 the government of Ontario made provision for the establishment of a provincial technical school, separate from the provincial university in Toronto. There was considerable ambiguity as to whether this was to be a school of applied science or of crafts. The former idea won out and in 1878 the School of Practical Science became closely associated with the University of Toronto. It enrolled seven students. At first it offered only a diploma but in 1884 the degree of civil engineer was established and in 1885 there was a single graduate. This experience may sound paltry but some of the earliest graduates of the Toronto programme went on to illustrious careers in the USA and did much to establish the *bona fides* of the programme.[25]

In 1874 Laval University, at its Montreal campus, established an *École polytechnique*. It began with 12 students and sent forth its first graduates in 1877. In the same year, tiny King's College in Windsor, Nova Scotia, began an engineering programme. Dalhousie University initiated a three-year diploma course in 1886. Not to be overlooked was the Royal Military College (RMC) at Kingston, Ontario. Established in 1876, it was primarily intended to provide scientific training to military officers yet, from the outset, it had the dual objective of producing civilian engineers. Of the 20 or so students per year it turned out up to 1890, the number going into civilian practice would have constituted a significant proportion of the graduate engineers in Canada.

It is evident, however, that prior to the early 1890s engineering education was almost a trifling matter in Canada. There was very little supply, and for that matter, not much demand. Apart from railway building it seems that not much was going on in the Canadian economy that called for scientifically trained personnel. As late as 1900 the dean of engineering at Toronto was suggesting that, with still only 10 graduates per year, his institution was producing more engineers than the economy was absorbing and that a large fraction of his graduates had to emigrate to the USA. The Canadian economy was evidently getting by with little scientific input and calling upon few engineers. The biggest need was probably for metallurgists in the new steel plants and for electrical engineers to plan the electrical generating stations and distribution systems that were beginning to be built.

Things appear to have begun to change, albeit still in a small way, in the early 1890s. New university programmes were established. The existing ones experienced large increases in enrolment. In 1892 engineering enrolment at McGill and Toronto, the two leading programmes, more than doubled and continued on an upward trend thereafter. The University of New Brunswick established an engineering programme that saw its first graduate in 1892. Queen's University at Kingston, Ontario opened an applied science programme, for financial reasons nominally as an independently organized School of Mines, and it produced its first graduates in 1897. Table 3.4 shows the annual number of engineering graduates in Canada from 1890 to 1914, not including graduates of the Royal Military College. The numbers jump in 1893 but still remain remarkably small until another jump early in the twentieth century. It was not until 1904 that Canadian universities were producing more than 100 newly-minted engineers each year.[26] The net contribution to the stock of scientifically educated engineers in Canada would have been smaller still. People in all walks of life were leaving Canada in large numbers and engineers would have been among them. Still, university

Table 3.4 Engineering graduates of Canadian universities, 1891–1914

Year	Total	McGill	Toronto	EP*	Queen's	Other
1891	14	11	–	3	–	–
1892	22	17	1	3	–	1
1893	41	18	10	7	–	6
1894	41	23	12	2	–	4
1895	43	30	11	–	–	2
1896	48	33	9	5	–	1
1897	50	41	5	–	2	2
1898	81	42	9	25	3	2
1899	65	41	6	14	1	3
1900	58	36	10	4	4	4
1901	86	40	20	22	3	1
1902	72	30	15	7	14	6
1903	96	41	16	17	18	4
1904	105	53	19	13	14	6
1905	117	43	24	16	21	13
1906	139	64	28	14	19	14
1907	169	62	31	28	35	13
1908	213	89	50	30	34	10
1909	229	82	67	31	43	6
1910	265	108	64	33	42	18
1911	294	100	99	43	42	10
1912	364	97	129	55	67	16
1913	354	106	106	63	59	20
1914	364	115	151	41	42	15
pre-1891	164	140	3	10	–	11
Total to 1914	3,300	1,322	892	476	433	177
Total to 1911	2,218	1,004	506	317	265	126

* EP is the École polytechnique.
Source: Hamis (1976).

engineering programmes were on a reasonably solid footing by the early years of the twentieth century and continued to grow. They might best be characterized as growing not ahead of national demand, or lagging much, but hand-in-hand with it. The 105 graduates of 1904 had doubled by 1908 and had increased by another 70 per cent by the time of the outbreak of the First World War.

Two universities, McGill and Toronto, dominated the engineering scene in Canada from the outset. McGill was the early leader. Toronto lagged considerably, catching up with McGill's output of engineering graduates only in 1908, but by 1914 the University of Toronto had surpassed McGill and was well established as the leading producer of engineers in the country. Two other university programs, the *École polytechnique* in Montreal, the only French language engineering programme in the country, and Queen's

University at Kingston, Ontario, made up the bulk of the residual.[27] At least on the surface this appears to be a thin basis upon which to build the foundation of engineering expertise on which the burgeoning Canadian economy depended. It would appear that Canada was getting by on a slender few engineers in the first few years of rapid economic change but may have been producing an adequate supply by the early years of the new century. By 1914 Canada's universities had turned out a cumulative total of 3,300 engineering graduates (still not counting the RMC). A not insignificant number of those would have emigrated and, as has always been the case with engineers, some would have abandoned the practice of engineering for a wide range of other pursuits, especially in business management and in public administration.

What the foregoing reveals is that Canada was not seriously failing to produce domestically educated engineers. University programmes were in place and were rapidly expanding their output. By 1911 a cumulative total of 2,218 engineers had graduated from Canadian universities. It appears that number can probably be fairly reconciled with the 1911 census count of 3,157 Canadian-born engineers (Canada, 1911 Census, vol. 6). There were still numerous engineers active in Canada who had been trained through the old apprentice system although that route to the profession was disappearing rapidly after 1900. The census of 1901 (Canada, 1901 Census) had recorded 2,076 professional engineers in the country. Up to that time only 667 had graduated from universities. That would imply 1,409 non-university trained engineers of the older generation. If 90 per cent of them survived to 1911 we might postulate that in that year there would have been 1,268 non-graduate engineers practising in Canada. By 1911 a cumulative total of 2,218 had graduated from Canada's universities. Some of those would have died and more of them would have emigrated: say, 22 per cent.[28] That would place the estimated stock of Canadian engineers at 2,998, and since there might still have been a few shop-trained engineers entering the system, that is a number that is remarkably close to the 3,157 Canadian-born engineers enumerated in 1911.

Too much emphasis should not be placed on that apparently close correspondence. Not all Canadian-born engineers would have been educated in Canada. Moreover, the graduates of Canadian university programmes would not all have been Canadian-born. Nevertheless, it is moderately reassuring that the numerical evidence we have at hand seems to be of the right order of magnitude. What is needed, however, is a richer body of evidence. We would like to know more about the nature of Canadian engineers: where they came from, how they were trained, and what role they were playing in the economy. The main point is that Canada was doing reasonably well in educating a cadre of engineers, and evidently did not lag far behind the USA in that regard.

Evidence from the *Dictionary of Canadian Biography*

The *Dictionary of Canadian Biography* (*DCB*) indexes its entries by categories, one of which is engineer.[29] That makes it fairly easy to access accounts of the careers of the small number of Canadians who for one reason or another were prominent enough to get a notice in the *DCB*. These biographical accounts can be instructive in indicating the kinds of things engineers were doing in Canada at around the turn of the twentieth century. Of course this is a select group, but not always chosen for their engineering accomplishments. Some were war heroes, others were prominent as public servants. Almost all were of an older generation since to get noticed in *DCB* they had to have died before 1920. One consequence of that is that their contributions to engineering and to the Canadian economy were mostly in connection with an older technology. These men, and of course they were all men, were from the generation of engineers who built the canals and the railways. Some gained prominence in the geological exploration of the country. Only a few represented the electrical and chemical technology of the new age.

Among those whose careers are outlined in *DCB* are the most prominent and famous of Canada's early engineers: Sandford Fleming, Casimir Gzowski, T.C. Keefer, Andrew Onderdonk and Thomas Willson. Except for Willson these were all men associated with the surveying and construction of Canada's railways. In the latter half of the nineteenth century that was high on the list of the nation's need for engineers. Willson, as previously mentioned, may be the outstanding case of Canadian contribution to the new technology of the Second Industrial Revolution. Raised in Hamilton and educated at the local high school, Willson moved to New York City to promote his ideas on uses of electricity. Through an attempt to produce aluminium electrolytically he discovered, and patented, a process for making calcium carbide. Willson's patents formed the foundation of what was to become the Union Carbide Corporation. Among the first uses of calcium carbide was the production of acetylene. It was initially thought to hold considerable promise as a lighting gas. Oxyacetylene welding was introduced a bit later, in 1903. Willson sold his US patent rights to Union Carbide, retaining only the Canadian rights, and returned to his native land. There he built and operated several plants to produce calcium carbide, making effective use of cheap hydroelectricity. He settled in Ottawa and continued an active career as an inventor and promoter of new industries. He had developed a low cost way of manufacturing nitrogenous solids that could be used as fertilizer. When an ambitious plan for a chemical, hydroelectric and wood pulp venture at Shipshaw, Quebec, faltered, Willson's assets were seized by his financier, James B. Duke, who went on to develop the site as an aluminium smelter. Willson started afresh but ran into problems getting financed as the First World War had just broken out. While canvassing Wall Street in 1915, Willson died from a heart attack at the prime age of 55.

A small collection of stories about highly selected engineers does not constitute much of a database. In the three volumes of *DCB* covering the period 1891 to 1920 there are 51 profiled engineers. Just under half (45 per cent) were born in Canada and 39 per cent were British-born. Interestingly, only four individuals were immigrants from the USA.[30] The *DCB* as a source of information is mainly useful for the light it can throw on the detailed nature of the careers of these notable engineers. It also permits us to know something about the more prominent of the earliest engineering practitioners in Canada. Besides the fact already mentioned that they were mostly involved with the laying-out of the railway system, we learn that only one-third of them had formal education in engineering or science. Many of them combined engineering with surveying or architectural practices, or had a range of business interests. They were not operating as full-time engineers. In the mid-nineteenth century large projects that required engineers were intermittent. It was common to find employment in the public service, at least for parts of a career, although that may be indicative of the selectivity of DCB. Nevertheless, it appears to be the case that in the latter half of the nineteenth century governments were more reliant upon engineering expertise than were businesses.

Brief examination of a few cases may help to give a sense of who were the engineers and what they did. William Tyndale Jennings, described at the time of his death as the dean of civil engineers in Canada, typifies the older group. He went from secondary education at Upper Canada College, the province of Ontario's most prestigious high school, to an apprenticeship with the Ontario Department of Public Works. After a short stint with the Great Western Railway he worked for Fleming on the survey of the route of the Canadian Pacific Railway. Jennings went on to become chief engineer for the city of Toronto but later took on projects all over North America as a consulting engineer. He acted, for example, as the chief engineer on the construction of the Crow's Nest Pass line of the Canadian Pacific Railway.

Job Abbott was an American of exquisite qualifications having attended Phillips Andover and the Lawrence Scientific School at Harvard. He came to Canada as an experienced practitioner, brought in as a consultant to the Toronto Bridge Company, which·had been induced into existence by the National Policy tariff in 1879. The following year Abbott was named president and chief engineer of the company, but in 1882 he moved to Montreal to form the Dominion Bridge Company. Abbott was the driving force behind the development of that prominent firm. In 1890 he moved back to the USA where, shortly thereafter, he died at a mere 51 years of age.

Charles Esplin was of Scottish birth but had moved to Canada with his family in 1846. He was an early student of engineering at McGill and established a business erecting grist mills and saw mills. In 1878 he moved to Winnipeg in the expectation of putting up mills during the Manitoba settlement boom. When that boom came to a screeching halt in 1883, Esplin moved to the USA, and while in the employ of a Minneapolis manufacturing company patented several improvements to milling machinery. He moved on to Seattle, to Victoria, then back to Winnipeg in 1897 as engineer to the Vulcan Iron Works. Esplin claimed to have set up Winnipeg's first electrical lighting plant during his earlier stay in that city.

Thomas Macfarlane was born in Scotland and there received a formal education in chemistry, capped by study at the prestigious Saxon Mining School in Freiburg. He moved to Canada in 1860 to be the manager of the Acton Copper Company in Quebec. Five years later he was engaged by the Geological Survey of Canada, and in 1868 discovered and developed Silver Islet in Lake Superior.[31] Later he became a mining consultant to Joseph Wharton at Bethlehem Steel and tried, without success, to interest Wharton in the Sudbury Basin, which later became prominent in the production of copper and nickel. In 1881 Macfarlane was a chemist and co-owner of a paint factory in Montreal. From 1886 onwards he spent the rest of his career as the chief chemical analyst for the Department of Inland Revenue and Customs.

The final example I shall give is Thomas Pringle, a largely self-taught millwright born in Lower Canada. He developed a particular interest in the exploitation of the hydraulic power provided by the Lachine Canal. It seems that he was responsible for installing two-thirds of the 76 turbines placed along the canal. Pringle operated as a consultant out of the Caledonia Iron Works at Lachine and seems archetypical of the pioneer type engineer in Canada. Nevertheless, Pringle was highly adaptive and became a pioneer in the use of hydraulic power to generate electricity. His Lachine Rapids Hydraulic and Land Company was among the first to use St Lawrence river water for that purpose. In 1892 he established T. Pringle & Son, the oldest full-scale firm of consulting engineers in the country. That firm designed the hydroelectric installations at Shawinigan Falls, at Chaudiere Falls south of Quebec City, and at the Long Sault in Ontario. Pringle was a charter member of the Canadian Society of Civil Engineers, formed in 1887.

These individual cases, and numerous others, serve to give something of the flavor of early engineering practice in Canada, and they point up the diversity of experiences to be found. They are far from adequate, however, to support generalizations about the supply of engineers, apart perhaps from showing that the country was capable of producing some prominent and successful engineers.

A sample of engineers practising in Canada in 1911

A published directory of engineers provides the material for a body of data on just over 400 engineers who were practising in Canada in 1911. *Who's Who in Engineering* was published in New York, and while it endeavoured to provide wide international coverage, it was very largely North American. Quite a large number of Canadians were included. Of the just over 18,000 entries, more than 15,000 of whom were engineers from the USA, 705 were engineers in Canada. For each person entered there is a significant amount of useful information. Typically, this included date and place of birth, educational history with an indication of where engineering skills were acquired, the branch of engineering pursued, work histories (commonly with specific dates), and notices of other accomplishments. A geographic index separately lists all those who worked in Canada with their specific places of residence or work at the time the data were assembled.

One might worry that a *Who's Who in Engineering* would be selective of an elite and far from representative of the general run of practitioners of the profession, but that does not appear to be the case. The main indicator of that is the large number of young people and recent graduates who are listed. Another is the abundance of engineers of modest situation from small towns and cities. I have no way at present to make formal tests of representativeness but there are no clear signals of alarm. All indications are that the coverage of the data is so diverse that this source should serve to give a good profile of the engineering profession in Canada.

A first edition of *Who's Who in Engineering* came out in 1922. I am working with the second edition, published in 1925, which is the earliest that I have available. This second edition is probably more suited to the task since the appearance of the first edition generated interest in the project and substantially increased the numbers and the range of responses to the next round. The data were assembled over the latter half of 1923 and the first months of 1924.

I have drawn from this listing all the engineers resident in Canada. The sample of interest to me, however, consists of those who were practising in Canada in 1911, towards the end of the period of especially rapid growth of the Canadian economy and the initial period of adoption of the new technology that typified the Second Industrial Revolution. To that end I have tabulated the records of 403 engineers whom the records show were active in 1911 and the information on residence and employment relates to that year. One implication is that an important source of bias would be that the data cover only those engineers who continued to be associated with the profession to 1924. Excluded would be those who died in the interval. As older members of the profession they would be more likely to have been trained by apprenticeship and not as likely to have had a university education. They would be more likely to have been born, and even trained, in Britain. Furthermore, they would be more likely to have been civil engineers than adherents to one of the newer branches of the profession. Also excluded would have been those who emigrated in the period between 1911 and 1924. There has always been a sizeable drain of qualified and ambitious Canadians to the USA, but the departure would also have included proportionally more of those who, in the first place, had migrated to Canada from the USA. American engineers quite commonly worked in Canada for periods before returning to their home country. To some unknown extent young American engineers did minor league service in Canada before returning to the majors. Finally there is that large group of people who are trained in engineering but who are drawn off into other lines of activity and who, by 1924, no longer thought of themselves or reported themselves as engineers. We need to be conscious of the possible effect these biases might have on any conclusions that may be reached.

One might also speculate that the directory might over-represent American and British engineers practising in Canada because they might have been more motivated to get notice in a directory that would bring them to the attention of their countrymen back home. If such a bias exists it works to my advantage since I shall argue that the proportions of immigrant engineers in 1911 were rather less than we might have been led to expect.

The data from the *Who's Who* sample can inform us on five variables of interest. The results, simply in terms of distributions of each of these five variables, are presented in Table 3.5. The first variable of interest is age. For convenience I have categorized the ages of engineers by four groups of birth dates. First there are the 'old timers', those born before 1865. They comprised only 9.7 per cent of the sample. It is worth remembering that the youngest of that group would have been only 47 years of age in 1911. Those no older than 36 years in 1911 made up 72 per cent of the sample. Clearly, engineers were predominantly a young lot.

A second variable of considerable interest is country of birth. Fully 70 per cent of the engineers in the sample were born in Canada. British-born entries comprised 16 per cent, and Americans 12 per cent. Canada was evidently not so dependent upon immigrant engineers as has commonly been presumed. It is also interesting that the immigrant engineers were more likely to have been British than American. To some extent that was an echo of the earlier immigration experience of the country. The British by birth in the sample outnumbered the British by training (the third variable), as we shall see shortly. The proportion of engineers born in the USA and those trained there are more closely balanced. The proportion of engineers who received their training in Canada was two percentage points above the pro-portion born in Canada. The proportion trained in Britain was 13 per cent and that trained in the USA about the same. There were a few instances of Canadians having received their training in Britain or the USA but the numbers are too small to support any generalizations. Some prominent Nova Scotia mining families sent their sons to the Royal School of Mines in London, and Harvard and Yale, with their Lawrence and Sheffield scientific schools respectively, drew a handful of Canadians. Overwhelmingly, though, Canadian-born engineers were trained in Canada, either by apprenticeship or in one of the small number of university engineering faculties.

Among the Canadian universities, McGill and Toronto, with 31 and 29 per cent of Canadian trained engineers respectively, dominated the

Table 3.5 Summary of evidence on engineers
in Canada in 1911

Period of birth Year	%
Pre-1865	9.7
1865–74	18.2
1875–85	49.1
1886 +	23.0

Country of birth Country	%
Canada	69.7
Britain	16.1
USA	11.9
Other	2.3

Training Country	%
Britain	12.9
USA	13.2
Canada	72.1
McGill	31.0
Toronto	29.3
Queen's	8.6
École polytechnique	8.3
RMC	4.8
Other universities	8.2
Non-university	9.7

Branch of engineering Type	%
Civil	47.0
Electrical	20.7
Mining	16.9
Mechanical	10.1
Forest	2.0
Other	3.3

Employment in 1911 Area	%
Private practice	19.0
Railways	16.7
Electrical utilities	9.7
Government	19.2
Other	35.4

Source: Tabulated from *Who's Who in Engineering*,
2nd edn (New York: Who's Who Publications, 1925).

national scene. The proportion of Toronto graduates corresponds reasonably with the university graduation records examined previously, but there is a notable shortfall of McGill products. Queen's graduates made up 8.6 per cent of the sample, and those of the *École polytechnique* in Montreal 8.3 per cent. The latter group is smaller than we would expect from the graduation numbers but it is not surprising that the *Who's Who* would draw fewer French Canadians. The Royal Military College provided the training of 4.8 per cent. That allows us to fill in a gap in the university graduation records and the number is quite plausible. All other Canadian university programmes together made up 8.2 per cent, leaving 9.7 per cent of engineers to report apprenticeships or other practical arrangements, which leads me to worry about a possible bias in the sample. Charles Mann, writing in 1918 in his *Study of Engineering Education* for the Carnegie Foundation and the National Engineering Societies (Mann 1918), claimed that, at that time, 'about half of the engineers in America were shop-trained, not school-educated'. Mann's reference total of engineers, from the US census of 1910, includes surveyors, and also electricians along with electrical engineers. The latter is a particularly egregious complication. It may well have led him to over-count engineers in his guess of 80,000. He reported (Mann 1918, p. 18) that membership in all engineering societies together amounted to 53,000, and that would probably have been an under-count of the total. Taking that latter number as the total, however, would still place 25 per cent of engineers in the USA as having been trained other than in universities and colleges. The much lower Canadian figure – just under 10 per cent – may reflect the younger age and greater recency of engineering training in Canada, or it may also be a reflection of a less extensive development of earlier manufacturing in Canada that would have meant fewer 'shops' to provide training.

With regard to branches of engineering (the fourth variable), it is not surprising that almost half (47 per cent) of Canadian engineers classified themselves as civils. What is most interesting is that the next most frequent type, with 20.7 per cent, was electrical engineering. Mining engineers comprised 16.9 per cent of the total, and mechanicals only 10.1 per cent. I cannot escape the suspicion that mechanical engineers may have been under-counted. The 1911 Census of Canada recorded a somewhat larger number of mechanical than of electrical engineers. That census, however, placed 30 per cent of all engineers in the 'branch not specified' category. Two hypotheses require further investigation, if appropriate data can ever be found. One is that *Who's Who in Engineering* systematically under-represented mechanical engineers relative to other types. The second hypothesis is that Canada may have had a sparse density of mechanical engineers, at least in comparison with the USA and possibly other industrial nations. There are indications from the histories of Canadian engineering schools to give a bit of support to this second hypothesis. At both Toronto and McGill mechanical engineering appears to have taken a back seat to other branches. Civil engineering was everywhere the largest programme, but electrical and mining received more attention than mechanical. There are also comments made by engineering deans that Canadian manufacturers were backward in appreciation of trained mechanical engineers. By contrast, firms had little hesitation in engaging electrical engineers when they came face to face with the electrification decision.

The prominence of mining engineers parallels the emphasis given to geology in Canadian science in the late nineteenth century. There seems to have been a perception that Canada had surely been amply endowed by nature; all that was required was to explore and discover. Great iron deposits were a foremost hope but it was not until well into the twentieth century that those would be found. A shortfall of mechanical engineering may represent an adaptation to the evident demand for engineers in Canada. At the same time it may have constituted a weak foundation for the development of a wider range of manufacturing industry.[32]

A final tabulation of Canadian engineers in 1911 categorizes them by the nature of their employment. Almost one-fifth were in private practice as engineering consultants. A similar proportion worked for governments. The railway companies were a large employer of engineers and we have to keep in mind that in 1911 Canada was going through a peak period of railway construction. The largest fraction of immigrant engineers were recent arrivals brought into the country to lay out new railway lines and to supervise their construction. These immigrant railway engineers were much more likely to have come from Britain than from the USA. The situation was accentuated in the years immediately following. In focusing the sample on engineers practising in Canada in 1911, I set aside registrants in the 1925 *Who's Who* who had graduated from college and entered engineering practice after 1911, and those who immigrated to Canada subsequent to that date. Almost all of the immigrants had come prior to the outbreak of the First World War. The number was large, amounting to 23 per cent of the number of immigrant engineers in my sample who were active in 1911. The records show that a large proportion of these immigrants were employed by the railways.

Almost 10 per cent of engineers worked for electrical utilities. This was a time when the country was feverishly electrifying; indeed, that was an outstanding feature of the Second Industrial Revolution. It shows up in the allocation of engineering talent. Not only were 10 per cent of engineers working for electrical utilities, but many of those in private consulting practices were involved with electrification as well.

The largest category of employment, with 35 per cent of all engineers, is a residual representing largely mining and manufacturing industries. No single subcategory of this residual appears to be large enough to tabulate separately. Many of these engineers were also involved with one or another aspect of electrification. Some were manufacturing electrical apparatus, while others were organizing production processes to make use of electricity. Electrolytic smelting was being applied to a whole range of mineral products. Remotely located pulp and paper mills were among the earliest users of electrical power. Canadian General Electric and Westinghouse of Canada stand out in the manufacturing sector as employers of engineers. Both of these companies appear to have relied quite heavily upon engineers brought in from the USA although they also employed senior engineers who were born and educated in Canada. In both cases these were employees whose work histories showed that they had spent some time with the parent company in the USA. The work histories of the engineers in this 'other employer' group reinforce my sense that there was a paucity of mechanical engineers in Canada. There were very few who reported being employed by the prominent machinery manufacturing firms, and there was not a single representative from the nascent automobile industry.

Much engineering was done on a project by project basis. Engineers were engaged to design and oversee the setting-up of new plants. Once those were completed they would move on to other projects. As early as the late nineteenth century there appears to have been a well-established industry in engineering consulting, augmented by numbers of highly mobile employed engineers. This was probably an effective way of allocating scarce scientifically skilled resources. In North America this industry operated on a continent-wide basis. It is an industry that has so far attracted little attention from scholars. Only by extensively examining business histories and business records will it come to be seen how important, quantitatively, was this aspect of the engineering trade. It is a reasonable guess that the Canadian economy was able to draw upon the full extent of consulting engineers in the USA to augment whatever demands could not be met by Canadian domiciled engineers. The employment histories related by the respondents to *Who's Who in Engineering* reveal considerable mobility both across employers and across localities.

Concluding remarks

Only very tentative conclusions can be reached from this survey of the pieces of evidence relating to engineering expertise in Canadian industrialization. This is not a topic on which there is much of an established literature. My study is more in the nature of a preliminary survey. Because Canada evidently succeeded so well in exploiting the technology of the Second Industrial Revolution, it should not be surprising that the evidence suggests that, overall, Canada was generating an adequate supply of engineers. That claim has to be qualified, however, by the fact that Canada's unusually rapid economic growth got underway in the late 1890s, before the domestic supply of graduate engineers had been much developed. Shortly thereafter, Canada's engineering schools were developing vigorously and a good supply of engineers was being domestically produced. In that earlier period, though, domestic university graduates must have been considerably supplemented, either by 'shop-trained' engineers or by purchased consulting services, primarily from the USA. We should also take into account the fact that it seems that a quite small number of engineers played key roles in the development of many of the new industries. For example, one man, Alexander Holley, was responsible for the design and construction of most of the Bessemer steel plants built in the USA (see McHugh 1980). A relatively small number of large hydroelectric installations dominated the Canadian scene. They may not have required many engineers to design and erect them and it should be possible to trace the key individuals involved. Larger numbers of engineers may have been required for the continued operation of the new technology than for the initial implementation.

Especially with regard to the new technology, Canada does not appear to have been acutely dependent upon immigrants. The country appears to have been doing particularly well in implementing the electrical technology and it was making good use of its forest and mineral resources. The one serious question mark relates to mechanical engineering and the provision of adequate support for mechanical types of manufacturing. An important point that has to be recognized is that engineering services were being traded internationally in an extensive way. We know that Canadian engineers were engaged on projects throughout the world, but especially in Latin America. Canadian engineers routinely consulted on projects in the USA, and that was especially the case with mining engineers who ranged freely around the globe on a project basis. What we do not yet know is the extent to which Canadian businesses were purchasing engineering services internationally, particularly from the USA. That is something that will have to be determined by evidence from the demand side of the market, although a survey of engineering consulting firms in the USA would also be helpful.

This chapter has focused especially on the supply of engineering expertise. It has brought together some data that have not been previously looked at in this context. It has pointed up an important issue in Canadian economic development that needs more attention. The main thing it records is that Canada appears to have done reasonably well in generating a supply of engineers. It would be helpful to gather information from the other side of the market, to look at individual businesses and industries in order to inquire into their sources of engineering expertise, and the extent of their needs for it. That will be a time-consuming enterprise of delving into business histories. I have done some of that for the principal industries, especially steel and pulp and paper, but Canada has not been abundantly served with published business histories and so I see a task that will be long and difficult. In the meantime, it should be helpful to have made a start at exploring the supply side. We are able to gain some perspective on the nature of the supply of engineers that Canada was able to assemble, which allows us to garner some sense of the complex detail that underlay the implementation of the fruitful new technology and to have some appreciation of how Canada was able to make such impressive productivity gains from it.

Notes

1 The long-appreciated fact that between 1897 and 1913 real per capita income grew more rapidly in Canada than in any other nation is sustained by Angus Maddison's most recent manipulations of the data (Maddison 1995). That conclusion would not seem to be altered by the alternative treatment of international comparability proposed by Leandro Prados, although Prados provides data only for 1890 and 1900 as well as 1913 (Prados de la Escosura 2000).

2 There is no thoroughgoing, up-to-date study of electrification in Canada. For one part of central Canada the topic is examined by John Dales (1957). Some aspects of the topic for the province of Ontario are covered by H. V. Nelles (1974). One aspect of the topic, dealing specifically with productivity, but mainly for a slightly later time period, is dealt with by Peter Wylie (1989).

3 The outstanding example is the chemically-based manufacture of paper from wood, but the inter-relation between chemical processes and cheap electricity was also significant. Examples are the electrolytic production of calcium carbide and the electrolytic smelting of non-ferrous metals.

4 A decade is an admittedly rather short period of time. One might wonder whether international comparisons may be sensitive to the precise period chosen. That is not the case. Comparisons over various periods do not alter the conclusion. An examination of the growth record of many countries shows that really rapid growth often occurs in spurts. It is not usually sustained over very long periods. In the Second Industrial era, broadly defined, no country outperformed Canada. Italy came closest to the Canadian growth performance. That country appears to have avoided the depression of 1907/8 and continued to grow rapidly through 1911. The peak period of Italian growth was between 1898 and 1911 when that country grew at an average rate of 4.19 per cent: closer to, but still not exceeding, the Canadian performance. No other country comes really close. The peak rate of growth for the USA at this time was 3.32 per cent per annum.

5 The data are from Mitchell (1998).

6 Industrialization in Canada can be examined in terms of GNP originating in what is effectively the two-digit industry level. The available numbers are in current dollars and there do not presently exist industry-specific price series to convert these to an index of industrial production. The numbers can, however, be deflated by the same aggregate price index as is used for GNP so as to remove the effects of change in the value of money. A comparable calculation made for the USA in the same period using data from Kendrick (1961) generates a rate of change of industrial production that is virtually identical to that reproduced in Mitchell (1998).

7 The elaboration of this argument is the main subject of McInnis (1999).

8 This is not to imply that I subscribe to Rostow's model of the 'take-off'. Gerschenkron, however, was probably on to something in emphasizing, as did Schumpeter, that economic growth comes in discernible spurts. Those early ideas about the growth process are readily absorbed into the emphasis given to determinate 'shocks' in the writings of the New Growth theorists.

9 It is most common to think of the extent of industrialization in terms of the structure of the economy: the proportion of total output accounted for by manufactured goods. It is quite true that in that sense Canada was still a predominantly agricultural economy. In 1890 manufacturing accounted for only 26 per cent of Canada's GDP while agriculture, fishing and forestry made up 31 per cent. It is

less common, but no less instructive, to look at GDP originating in manufacturing per person in the country. It is by that standard that I state that Canada was already a highly industrialized country.

10 One should not think of these industries as unaffected by the technology of the Second Industrial Revolution. Cotton textile factories in Canada, for example, were among the first to install electric lighting. Open-flame lighting had been a particular hazard in cotton mills. Printing establishments were among the first to use electric motors to drive their machinery. Light, battery-driven motors could be used with printing machinery, and the publishing industry was in the forefront of electrification, well before the end of the nineteenth century. The newspaper in St Catharines, Ontario was the second printing plant in all of North America to use electric motors (Biggar 1920, p. 32).

11 They are not, it should be emphasized, simply the mechanical result of starting from a small base. The iron and steel industry contributed 21.71 per cent of manufacturing output by 1897, and the non-ferrous metal industry 5.56 per cent. Electrical apparatus manufacturing and rubber products were, indeed, small new industries but they did not grow notably faster than the larger, established industries.

12 This latter point is consistent with the findings of Broadberry (1997), relating to productivity differences between the USA, Britain and Germany.

13 This is a transfer of technology because Canadians contributed little in the way of inventions to this new technology. What might be regarded as the very leading edge of the technology of the Second Industrial Revolution had been contributed by a Canadian. Nova Scotian Abraham Gesner had found a use for crude petroleum when he patented the illuminating oil that he called kerosene and thereby kicked off the development of the world petroleum industry. That was in 1855, and by the late nineteenth century petroleum and its products were a long-established part of the economy. While it has many of the characteristics of Second Industrial Revolution technology, kerosene was part of an earlier age. No comparable invention was made by a Canadian to contribute to the new technology. The closest contender might be the development by Thomas Willson of the process for synthesizing calcium carbide.

14 The standard reference on the steel development at Hamilton is Kilbourn (1960). See also Donald (1915).

15 The development of the Algoma Steel Corporation is covered by McDowall (1984).

16 This turned out to be chimerical. The Helen Mine near Wawa, Ontario, proved to have intractable ore.

17 Within a very few years the Algoma Steel Corporation entered its first of several bankruptcies. In late 2001 it was being rescued once again by large subsidies from both the Province of Ontario and the government of Canada.

18 Electrical apparatus manufacturing also involved much American entrepreneurial direction. Both General Electric and Westinghouse played a large role in the Canadian industry through subsidiary plants. Smaller firms, such as the Packard Electric Company that made transformers, were also American-owned. Some of the earliest developments, however, were made by indigenous Canadian firms which were later absorbed by the large American concerns.

19 I am convinced that their adjustment, based on the relative numbers of civil engineers and surveyors in 1930, is surely too small. The proportion presumed to be surveyors is much smaller than in Canada in 1911 where the census separately tabulated surveyors and civil engineers. On the other hand, it is unclear what Blank and Stigler (1957) did about electrical engineers and they may not enter

their total at all. The two miscounts would be to some degree offsetting so their estimate of the total number of engineers may not be too far off the mark.

20 Benjamin Chaffey of Brockville, Ontario was active around the middle of the nineteenth century as an architect, engineer, contractor and manufacturer. He supervised the construction of canals on the St Lawrence River, operated a factory that manufactured lock gates for canals, contracted a section of the Grand Trunk Railway of Canada, and designed and built many of the stone piers upon which the Grand Trunk's Victoria Bridge at Montreal rested. To carry out that last project he designed a unique travelling crane for which he gained considerable fame. His engineering skills were entirely self-taught. His nephew, George Chaffey Jr of Kingston, Ontario was largely self-taught as well, although inspired by his uncle. George Jr designed and built steamships for the Great Lakes, for which he gained notice in the *Scientific American*, before going on to lay out irrigation projects in California and Australia and, in 1884, to found the Los Angeles Electric Company.

21 Sir Sandford Fleming was a Scottish trained immigrant who gained prominence as the principal surveyor and chief engineer of the Intercolonial Railway of Canada. He is also well known for his international campaign for standard time zones. In the late nineteenth and early twentieth centuries he was probably Canada's best known engineer. Sir Casimir Gzowski was an engineer and military officer exiled from Poland for his role in an insurrection. He was an early promoter of railway development and carried out the construction of a part of the Grand Trunk Railway of Canada.

22 The literature on the development of engineering education in Canada is sparse. I have been aided considerably by the unpublished PhD dissertation of Mario Creet (1992). The published histories of the individual institutions are scanty on the development of engineering. Young (1958) on the University of Toronto is more substantial than most of what is available. Harris (1976) has a succinct chapter on engineering and, fortunately, has tabulated the annual number of graduates of all the universities with engineering programmes.

23 At about the same time King's College in Fredericton, shortly to become the University of New Brunswick, offered instruction in civil engineering under the guidance of a recently arrived British engineer. Only a tiny handful of students pursued the programme.

24 The 1870s was a prime decade for the establishment of university engineering programmes in the USA. It appears that Canada did not lag notably in starting university programmes in engineering. The numbers, however, were small.

25 E. W. Stern was one of the very earliest graduates of Toronto's engineering school. He and Kennard Thomson (class of 1886) became renowned for their foundation designs for large New York buildings.

26 At that time, according to Ahlström (1982, p. 107), Germany was producing about 4,400 new engineers per year and France more than 1,400. Sweden, a country slightly smaller in population than Canada, was turning out about 175. In the USA, the institutions of New York state alone were producing more than 400 engineers per year (Edelstein 2001, Table 10).

27 In Sweden, however, which produced more engineering graduates than Canada at the time, there were only two schools.

28 That proportion is based on the ratio of Canadian-born engineers living in the USA in 1910 to the number in Canada at the time. The calculated ratio is 19.7 per cent but electrical engineers could not be taken into account and they were younger and somewhat more likely to have emigrated. A figure of 22 per cent seems reasonable.

29 The *Dictionary of Canadian Biography* (Francess G. Halpenny and Jean Hamelin, general editors) is an on-going project in several volumes. Volume XII, covering persons who died between 1891 and 1900, was published in 1990.

30 It should be clarified that one did not have to die in Canada to be included in *DCB*. Two of four Americans returned to the USA before they died, but they had engineering careers in Canada.

31 Silver Islet was one of Canada's first consequential metal mines. It was a tiny island of about fifty metres in each direction from which a very large amount of silver was extracted. An American engineer was brought in to tackle the formidable task of developing the site and the mine workings under the severe conditions encountered.

32 One might turn this into a path dependency type of argument. Adapting to the manifest need for electrical and mining engineers, Canadians may have left themselves deficient in the mechanical engineering expertise that would have allowed them to take fuller and wider advantage of all aspects of the new technology. Some commentators have long bemoaned a lack of greater development of machinery manufacturing industry in Canada. Whether or not that is a valid concern is arguable but we may here have a glimpse of one reason.

References

Ahlström, G. (1982), *Engineers and Industrial Growth* (London: Croom Helm).

Biggar, E. B. (1920), *Hydro-Electric Development in Ontario* (Toronto: Biggar Press).

Blank, D .M. and G. J. Stigler (1957), *The Demand and Supply of Scientific Personnel* (New York: National Bureau of Economic Research).

Broadberry, S. N. (1997), *The Productivity Race: British Manufacturing in International Perspective, 1850–1990* (New York: Cambridge University Press).

Canada, Department of Agriculture, 1901 Census of Canada (1910), 'Occupations of the People' , *1901 Census Bulletin*, Vol. XI (Ottawa: King's Printer).

Canada, Department of Trade and Commerce, 1911 Census of Canada (1915), 'Occupations of the People', *Fifth Census of Canada*, Vol. VI (Ottawa: King's Printer).

Creet, M. (1992), 'Science and Engineering at McGill and Queen's Universities and at the University of Toronto, 1880s to 1920s' (Kingston, Ontario: unpublished Queen's University PhD thesis).

Dales, J. (1957), *Hydroelectricity and Industrial Development in Quebec, 1898–1940* (Cambridge, MA: Harvard University Press).

Donald, W. J. A. (1915), *The Canadian Iron and Steel Industry* (Boston, MA: Houghton Mifflin).

Easterbrook, W. T. and H. G. J. Aitken (1965), *Canadian Economic History* (Toronto: Macmillan), 2nd edn.

Edelstein, M. (1988), 'Professional Engineers in Australia: Institutional Response in a Developing Economy, 1860–1980', *Australian Economic History Review*, 28, pp. 8–32.

Edelstein, M. (2001), 'The Production of Engineers in New York Colleges and Universities, 1800–1950', paper presented to the Rochester Conference in honour of Stanley Engerman, 8–10 June 2001.

Fowke, V. C. (1957), *The National Policy and the Wheat Economy* (Toronto: University of Toronto Press).

Halpenny, F. G. and J. Hamelin (general eds), *Dictionary of Canadian Biography* (Toronto: University of Toronto Press, various dates).

Harris, R. S. (1976), *A History of Higher Education in Canada, 1663–1960* (Toronto: University of Toronto Press).

Keay, I. (2000), 'Canadian Manufacturers' Relative Productivity Performance, 1907–1990', *Canadian Journal of Economics*, 33, pp. 1,049–68.

Kendrick, J. W. (1961), *Productivity in the United States* (New York: National Bureau of Economic Research).

Kilbourn, W. (1960), *The Elements Combined: A History of the Steel Company of Canada* (Toronto: Clark, Irwin & Co.)

Maddison, A. (1995), *Monitoring the World Economy, 1820–1992* (Paris: OECD).

Mann, C. (1918), *A Study of Engineering Education* (New York: National Engineering Societies).

McDowall, D. (1984), *Steel at the Sault: Sir James Dunn and the Algoma Steel Corporation 1961–1956* (Toronto: Toronto University Press)

McHugh, J. (1980), *Alexander Holley and the Makers of Steel* (Baltimore, MA: Johns Hopkins University Press).

McInnis, M. (1994) 'Immigration and Emigration: Canada in the Late Nineteenth Century', Ch. 7 of T. J. Hatton and J. G. Williamson (eds), *Migration and the International Labor Market, 1850–1939* (London: Routledge).

McInnis, M. (1999), 'Canadian Economic Development in the Wheat Boom Era: A Reassessment', unpublished paper presented to the annual meeting of the Canadian Economics Association, Toronto, May 1999.

Mitchell, B. (1998), *International Historical Statistics: Europe*, 4th edn (London: Macmillan).

Nelles, H. V. (1974), *The Politics of Development: Forests, Mines and Hydroelectric Power in Ontario, 1849–1941* (Toronto: Macmillan of Canada).

Prados de la Escosura, L. (2000), 'International Comparisons of Real Product, 1820–1990: An Alternative Data Set', *Explorations in Economic History*, 37, pp. 1–41.

Rosenberg, N. (1998), 'Chemical Engineering as a General Purpose Technology', in E. Helpman (ed.), *General Purpose Technologies and Economic Growth* (Cambridge, MA: MIT Press), ch. 7.

Urquhart, M. C. (1986), 'New Estimates of Gross National Product, Canada, 1870–1926: Some implications for Canadian Development', in S. L. Engerman and R. E. Gallman (eds), *Long-Term Factors in American Economic Growth*, Studies in Income and Wealth, Vol. 51 (Chicago: University of Chicago Press).

Urquhart, M. C. (1993), *Gross National Product, Canada, 1870–1926: The Derivation of the Estimates* (Kingston and Montreal: McGill-Queen's University Press).

Who's Who in Engineering (1925), 2nd edn (New York: Who's Who Press).

Wylie, P. (1989), 'Technological Adaptation in Canadian Manufacturing, 1900–29', *Journal of Economic History*, 49, pp. 569–91.

Young, C. R. (1958), *Early Engineering Education at Toronto, 1851–1919* (Toronto: University of Toronto Press).

4
Technology Shifts, Industrial Dynamics and Labour Market Institutions in Sweden, 1920–95[1]

Lars Svensson

Introduction

The aim of this chapter is to explore the role of labour market institutions in promoting economic growth, notably by improving the compatibility between technology and human capital. The study draws upon elements from three traditions in social science: a cyclical model of economic growth developed within a Schumpeterian structural-economic framework is combined with a theory of skill-biased technological change from mainstream labour economics and a Northian approach to institutional change. The basic idea is that technological change generates divergent biases in demand for skills in different phases of a structural cycle. These divergent tendencies are the result of investments in some periods being directed mainly towards the renewal of products and processes, and in others towards increasing the efficiency of the established structure. Shifts in skill bias in turn create a pressure to transform labour market institutions.

Taking a long-term perspective, the history of the industrialized world has been characterized by a temporal pattern of alternation between periods of convergence and periods of divergence. In the words of Moses Abramovitz, there were periods when economically-strong regions and sectors were able to 'forge ahead', and other periods when lagging regions and sectors managed to 'catch up' (Abramovitz 1986).

Structural-economic analysis has demonstrated that variations between convergence and divergence in the global economy have been systematic and related to long-term structural shifts of a cyclical character (Schön 2003). From a Schumpeterian perspective, technological change has caused recurrent historical discontinuities and created new complementarities between central innovations and entrepreneurship, infrastructure, knowledge and institutions. To use the terminology introduced by Erik Dahmén (1950; 1988), investments in comprehensive development blocs have supported the new complementarities. Within the framework developed by Chris Freeman and Carlota Perez, successive technological paradigms are thought to be formed around major innovations, notably within the areas of power generation and communication (Freeman and Louca 2001; Perez 2002). Within a similar Schumpeterian setting, and on the basis of the comparatively ample historical data that are available for the Swedish economy, Lennart Schön has elaborated a cyclical model for Swedish economic development since early industrialization (Schön 1998; 2000): recurrent cycles are demarcated by structural crises at their endpoints. Between the crises, a cycle moves from an initial phase of transformation and renewal to a phase of rationalization. Empirical observations on a number of variables have revealed a pattern of development that has been generalized to form a reference cycle of roughly 40 years in length.

Transformation means '*changes of industrial structures*, where resources are reallocated between industries, and diffusion of basic innovations within industry... provides new bases for such reallocation' (Schön 1998, p. 399). This implies a change in the direction of economic and industrial growth as new products and production techniques are introduced. Eventually, as a result of continued adaptation and diffusion, increasing commodity supply and decreasing elasticity of factor supply give rise to increased competition in all markets. The resulting profit squeeze prompts a slowdown in investment activities, and a redirection towards short-term, efficiency-increasing investments; in other words, the economy enters the rationalization phase. This means '*concentration of resources* to the most productive units within the branches and measures to increase efficiency in the different lines of production' (Schön 1998, p. 399). As the limits of expansion within the established structure are reached, a structural crisis marks the transition to the next cycle. Three full cycles (approximately 1850–90, 1890–1933 and 1933–75), and an uncompleted cycle begun in the late 1970s, have been identified (Schön 1998). The transformation phase generally makes up half of each cycle, and is followed by a rationalization phase of between 15 and 20 years. There are strong indications that convergence in growth rates between countries, regions and sectors has coincided with phases of rationalization, while divergence has been dominant during transformation.

This observation relates to the explanations for the lack of convergence in some periods that have been advanced by Moses Abramovitz (1986). According to Abramovitz, one factor that may explain a follower's inability to catch up with the leaders is its lack of *social capability*. Although this is a rather broad and imprecise concept, the general idea is that variables such as level of education, institutional setting and vested interests configure the ability of an economy to adopt the most productive technology and organization of work. Since phases of transformation are characterized by the introduction of new technologies, social capability becomes a more decisive factor during transformation than during rationalization, when previously introduced technologies have been simplified and standardized and are consequently easier to adopt.

From a growth theory point of view, one could say that the concept of social capability incorporates many of the knowledge and human capital factors that have moved growth models from decreasing to constant or increasing returns. As previously suggested, periods of convergence conform to the expectations of the traditional growth models, while periods of divergence are more in line with 'new growth theory', as the growth effects of knowledge and human capital are substantially larger during transformation than during rationalization.

The relationships between physical capital, skills and labour are critical to the process of growth and transformation. New technologies can only be adopted and efficiently utilized by a labour force with proper skills. This means that demand for labour must be understood as demand for labour with qualifications that are compatible with the technology currently in use, or about to be introduced. This technology-specific labour demand is matched with supply in the labour market. Thus, the ability of labour market institutions to contribute to the matching process is instrumental to the social capability to adopt new technologies.

Two qualifications of this statement must be made. First, the introduction of new technology does not necessarily imply the utilization of more complicated modes of production. In the rationalization phase, it instead signifies the standardization of products and the simplification of production techniques. Second, new technologies are rarely introduced evenly across sectors and industrial branches. There are leaders in the process, and there are followers and laggards. Their interests with respect to labour market institutions are likely to differ.

Skill-biased technological change

The reasoning of the previous section resonates with current theoretical developments in labour economics. Several recent studies by labour economists have suggested that technological change is skill-biased, and that complementarity between capital and high-skilled labour is a crucial factor in the increase in the skill premium that has been observed in a number of countries during the past two decades (Berman, Bound and Machin 1998).[2]

The results of this research refer mainly to labour market conditions from the early 1980s onwards, the era of the electronic and biotechnological revolutions; but do they apply to other periods as well? Claudia Goldin and Lawrence Katz (1998) have attempted to trace the origins of capital – skill complementarity in the industrial past. Referring to the theory that the major technological advances of the nineteenth century substituted physical capital, raw materials and unskilled labour for highly-skilled artisans, they pose the following question: 'If, as the literature now claims, technological advance and human skill were not always relative complements, when did they become so?' (Goldin and Katz 1998, p. 694). The Goldin and Katz article suggests that, in the USA, the 1909–19 period exhibited the same sizeable complementarities between physical capital and high skills as we have seen in recent times. The authors conclude that a shift to capital – skill complementarity occurred along with fundamental changes in the production process in American industry in the early twentieth century; these included changes associated with continuous-process and batch methods, and the adoption of electric motors (1998, p. 716). Goldin and Katz imply that there was a 'once-and-for-all' shift from capital – skill substitutability to complementarity before 1920, which was later reinforced by further introductions of new technology, particularly from the early 1980s. Both periods bear the characteristics of a transformation phase.

The notion that complementarity between capital and labour is stronger during transformation than during rationalization fits our structural-economic framework of analysis, since radical changes in technology are concentrated during phases of transformation. The logic behind capital – skill complementarity is that investments in a new generation of capital open up possibilities for new combinations of resources. Success is closely related to the technical and organizational skills of the labour force. In other words, disparity in productivity due to differences in skills and ingenuity is greater the more radical is the change in technology.

During rationalization, the dominant direction of investment is towards the standardization of products and the simplification of production processes. Productivity gains from investment in physical capital and the organization of production tend to exceed the gains from investment in human capital. This outcome will reduce demand for high-skilled labour. Given that the alternation between transformation and rationalization is a recurrent phenomenon, however, we should not expect a once-and-for-all shift to complementarity, as suggested by Goldin and Katz, but rather cyclical variations in the skill-bias of technology.

Further, if we believe that wages reflect marginal productivity, this should be manifested in a cyclical pattern of change in the wage structure. The skill premium or the return to human capital investments should, *ceteris paribus*, be greater during transformation than rationalization. From a perspective of the firm, this means that there is a struggle for skilled labour, which has to be fought by means of competitive wages. The labour allocation function of the wage structure becomes prominent. The more advanced the technology, and the more promising in terms of potential profits, the higher the bid in the market. This implies that wages, as well as employment of highly-skilled labour, should rise above the average in leading firms and sectors.

During rationalization we expect a shift in demand to low-skilled labour. With the standardization and simplification of production, productivity gains from additional skills will be small. Because of the certain stickiness of wage structures a shift to relatively low-skilled and low-paid labour will be profitable. Moreover, the fierce competition and profit-squeezing that is typical of the rationalization phase make the cost-reducing function of the wage structure more important than the labour-allocating function.

Labour market institutions

In the labour market, prices and quantities (that is, wages and employment) are determined according to the principle of supply meets demand under certain institutional constraints. This formulation implies that wage formation, employment, and allocation of labour are neither entirely free market processes nor totally governed by institutions.

In economic analysis of labour markets, institutions are often regarded as static and exogenous. Within the framework developed in this chapter, however, labour market institutions are regarded as being under constant endogenous pressure to change. The rationale for this is twofold. First, our concern is with labour of varying quality in terms of skill. The cyclical approach has led us to hypothesize that demand for high-skilled labour is more pronounced in some periods than in others. Moreover, since the technological shifts that determine the structure of labour demand tend to be more pronounced in leading sectors, these sectors will demand relatively more skilled labour during transformation than will followers and laggards. In a similar way, during rationalization, leading sectors will demand relatively more low-skilled labour, because measures to standardize and simplify production are being emphasized. Accordingly, we may assume first that firms in leading sectors have a vested interest in forming institutions that support the supply of high-skilled labour during transformation, and of low-skilled labour during rationalization. We may assume also that firms in leading sectors have a different view of labour market institutions than followers and laggards do.

Second, following North (1990), we regard institutions as either promoting or impeding economic growth: in this case, by promoting or impeding technology shifts. Since new technology differs in character across the structural cycle, the demands that institutions have to fulfil in order to promote the adoption of new technology will also differ. Institutions that promote adoption under one set of circumstances may impede change under another. Thus, even if endeavours to form institutions have been successful, pressure for change may remain constant. Moreover, we may assume that leading sector firms will be the prime initiators of institutional adjustment.

An analysis of institutional change must consider both formal and informal institutions. The labour market exhibits a variety of informal institutions that together form a set of existing norms, hegemonic ideologies and wage policies. Changes of informal institutions occur in a different way, and at a different pace, compared with changes in formal institutions. Notwithstanding this, both types of change should be treated at the same analytical level since ideological conceptions, policy formulations, meta-agreements and material agreements (that is, over wages and employment conditions) contribute to a coherent institutional setting; that is, to a labour market regime. By this chain of events, the selection of informal institutions materializes into formal institutions. This process takes place within another set of institutions that regulates the inner life of labour market organizations (that is, workers' and employers' associations), and reflects power relations between opposing subgroups within the associations. An example of this is the degree to which power is centralized within these organizations.

Summary: a framework for the analysis of technology shifts and institutional change

The framework of analysis that has been developed in the previous sections can be summarized in the following sequence of events: technology shifts are the determinants of change in relative labour demand, which trigger institutional change, and thus create new equilibria in the labour market (see Figure 4.1). Changes in the structure of labour demand also influence equilibrium in the labour market through their effect on market forces. In our limited framework, technological change is treated as an exogenous factor.

Empirical analysis

First, we shall use the framework developed in the previous section to generate several hypotheses that will be discussed in the light of some empirical evidence. Our analysis will begin at the end of the chain in the labour market equilibrium link. New equilibria can be observed as changes in wage and employment relations between skill groups. Relative demand for high-skilled labour tends to increase during transformation and decrease during rationalization because of the specific character of new technology in each phase. Thus, we hypothesize that:

(a)   wages between skill groups tend to diverge during transformation and to converge during rationalization;
(b)   this tendency is most pronounced in leading sectors;

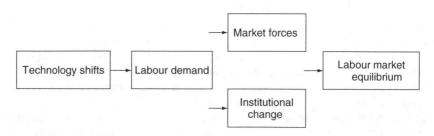

Figure 4.1   Technology shifts and the labour market

(c) relative employment of high-skilled labour accelerates during trans-
     formation and decelerates during rationalization;
(d) this tendency is most pronounced in leading sectors.

A specific labour market regime is compatible with specific labour demand
conditions. Since new technologies create new demand conditions, the
labour market regime will either promote or impede the abilities of firms
to meet these demands. Institutions that promote the adoption of one type
of innovation may impede the adoption of another type. Consequently, as
the industrial dynamic moves between transformation and rationalization,
there will be pressure on institutions to change. We further hypothesize that:

(e) institutions that promote wage convergence tend to be implemented
     during rationalization, and institutions that promote wage divergence
     tend to be implemented during transformation;
(f) this process is most prominent in leading sectors;
(g) leading sector agents are the most active in the process of labour market
     institutional change.

Both sets of hypotheses presuppose the identification of leading sectors in
the process of economic change. The proper approach would be to determine
theoretically a set of indicators that could be applied to sector-specific data
for different periods. This will have to wait until a later stage of this work,
however. At this stage, we have drawn upon the work of Chris Freeman and
others which refers to conditions in the industrialized world in general, and
which has been summarized in Freeman and Louca (2001, Table II.1). For our
period of investigation, the authors have identified the engineering industry
(and its subsectors), together with the chemical industry, as the leading
sectors until the mid-1970s, when they were replaced by the information
and communications technology (ICT) and biotechnology industries. This
is quite similar to the picture of the Swedish economy that emerges from
quantitative information on production and export performance, as well as
qualitative information on technological and organizational development
(Ljungberg 1990, chs 9–10; Schön 2000, chs 5–6). The identification of leading sectors can be summarized as follows:

Until the mid-1970s:
    chemical industry
    electro-technical industry
    mechanical engineering industry
        automotive industry (from the early 1950s)
        aircraft industry (from the early 1960s)
From the mid-1970s:
    ICT industry
    biotechnological industry (including the pharmaceutical industry)

Wage-structure shifts

The wage structure of the Swedish industry has experienced a secular development of great compression. This can be observed in a number of variables. Between 1917 and 1995, the gender wage-gap among blue-collar workers in manufacturing shrank from 42 per cent to 10 per cent. The decrease in the wage-gap for salaried employees in manufacturing was of roughly the same magnitude, although the differential was greater in this group. The ratio of male blue-collar wages to male white-collar wages in manufacturing rose from 45 per cent to 80 per cent during the same period.

This long-term wage compression has been characterized by alternation between periods of rapid equalization and periods of comparative stability, or even increasing wage differentials (see Figure 4.2). A striking feature is the apparent reciprocal similarity between the patterns of change in all three variables presented in Figure 4.2. This becomes even more evident in Figure 4.3 in which the deviations from a linear trend in the variables has been plotted. It is true that the 1920s exhibit a mixed picture, but this may be explained by the fact that the post-1918 period was one of extreme monetary turbulence. Excessive post-war inflation was replaced by sudden deflation

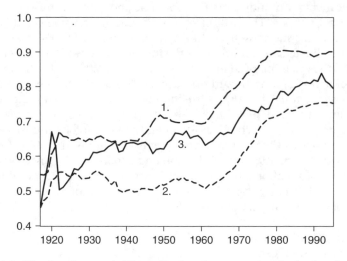

Figure 4.2   The female-to-male blue-collar hourly wage ratio (1), the female-to-male white-collar wage ratio (2), and the ratio between hourly wages of blue-collar workers and white collar workers (3), in the manufacturing industry

Note: Hourly wages for white-collar workers are calculated as the ratio of the average monthly salary to the average number of hours worked per month. Working hours are computed on the basis of information from Tegle (1982) and SOU (1976), p. 34.
Sources: My computations from *Sociala meddelanden (Social Reports)* 1917–27; *Lönestatistisk Årsbok (Statistical Yearbook of Wages)*, 1928–1951, SOS Löner *(Official Statistics of Sweden, Wages)* 1952–95.

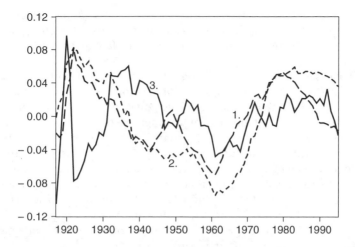

Figure 4.3 Deviations from the linear trend in the female-to-male blue-collar wage ratio (1), the female-to-male white-collar wage ratio (2), and the ratio between hourly wages of blue-collar workers and white-collar workers (3), in the manufacturing industry

Sources: My computations from *Sociala meddelanden (Social Reports)* 1917–27; *Lönestatistisk Årsbok (Statistical Yearbook of Wages)* 1928–51; *SOS Löner (Official Statistics of Sweden, Wages)*, 1952–95.

late in 1920, a shift that affected industrial sectors differently. Bearing this in mind, it can be stated that the period from the end of the war to the Great Depression brought a significant general decrease in wage differentials. Thereafter, all of our wage structure variables deviated negatively from the long-term trend through the 1950s, apart from a period after the Second World War. During the 1960s and the 1970s the wage structure compressed considerably before reaching the present level of equalization.

Since women on average possess lower labour market skills than men in each group, we may regard all of our wage structure variables as measures of wage differentials between labour at different levels of skill. They show a common pattern of long-term change and medium-term variation. In sum, the medium-term variations around the long-term trends of equalization can be structured roughly into periods as follows:

(a) from the end of the First World War until the Great Depression (compression);
(b) from the Great Depression until around 1960 (stability or increased differentials);
(c) from around 1960 until around 1980 (compression);
(d) from around 1980 onwards (stability or increased differentials).

These results clearly support our hypothesis that wages tend to diverge during transformation and converge during rationalization.[3] The theory that this tendency is most pronounced in leading sectors can be discussed only briefly and, owing to a lack of sector-specific wage data for non-production workers, must be based on the development of female-to-male blue collar wage ratios only.

We have selected the engineering industry as the leading sector until the mid-1970s. Figure 4.4 permits a comparison between the engineering industry and total manufacturing. Wage equalization after the First World War started earlier, extended over a longer period, and was more pronounced in the engineering industry than in total manufacturing. Likewise, the upward trend in the relative wage around 1960 appeared earlier in engineering, and the change was more pronounced than in total manufacturing.

To emphasize this point, the differences between relevant breakpoints in the two series are presented in Table 4.1. We chose the end of the First World War as the starting point, and the end year of our time series (1995) as the end point. Breakpoints were chosen in proximity to the shifts between phases of rationalization and transformation.

Although our analysis is based on only one wage structure variable, it indicates that wage-structure shifts were more pronounced, and appeared earlier, in the engineering industry than in total manufacturing, except for the last breakpoint. By this point, the engineering industry had ceased to be a leading sector. The industries that emerged as the leading sectors (ICT

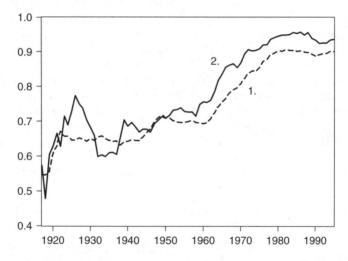

Figure 4.4   Female-to-male relative wages in total manufacturing (1), and in the engineering industry (2), 1917–1995

Sources: My computations from *Sociala meddelanden (Social Reports)* 1917–27; *Lönestatistisk Årsbok (Statistical Yearbook of Wages)* 1928–51, Socialstyrelsen, Stockholm; *SOS Löner (Official Statistics of Sweden, Wages)*, Socialstyrelsen, Stockholm 1952–60; Statistiska Centralbyrån, Stockholm 1961–95.

Table 4.1   Change in the female-to-male wage ratio of blue-collar workers in manufacturing and in the engineering industry (percentage points)

Manufacturing		Engineering	
1918–33	11.1	1918–30	20.4
1933–60	3.6	1930–58	−5.9
1960–82	21.2	1958–82	23.5
1982–95	−0.7	1982–95	0.0

Sources: My computations from *Sociala meddelanden (Social Reports)* 1917–27; *Lönestatistisk Årsbok (Statistical Yearbook of Wages)* 1928–51; *SOS Löner (Official Statistics of Sweden, Wages)*, 1952–95.

and biotechnology) were established only during the last part of our period. Consequently, there is no breakpoint in the wage ratios of these sectors.

Employment structure shifts

The renewal of Swedish industry after the structural crisis of the early 1930s was based mainly upon development blocs formed around two major innovations: the electric motor and the combustion engine. The transformation of the industrial structure brought relative growth to the engineering industry. Between 1935 and 1960, the engineering industry almost doubled its share of total industrial employment from 16 per cent to 29 per cent. During the following decade only a minor increase occurred, from 29 per cent to 30 per cent.

The share of non-production workers was considerably larger in the engineering sector than in other parts of the manufacturing industry. In 1935, white-collar workers constituted 16 per cent of the work force in engineering compared to 11 per cent in total manufacturing. In ASEA, one of the largest firms in the electro-technical industry,[4] the share of salaried employees amounted to 27 per cent.

As shown in Table 4.2, relative employment of non-production workers increased in all industrial sectors during the period from 1920 to 1996. Change was particularly rapid during the 1935–60 and 1975–96 periods (that is, during phases of transformation and renewal). During the rationalization phases of the 1920s and from 1960 to 1975, change was much slower. These results support our hypothesis that relative employment of skilled labour accelerates during transformation and decelerates during rationalization.

Table 4.2 also demonstrates that the temporal patterns of employment-structure shifts were similar in total manufacturing, the engineering industry and the chemical industry, although at different levels of relative employment. The observed increase in relative white-collar employment in total

Table 4.2A Proportion of white-collar workers (percentage of the total work force) in the manufacturing industry, the engineering industry, and the chemical industry

	Manufacturing	**Engineering**	**Chemicals**
1920	10	15	15
1935	11	16	19
1950	18	22	28
1960	24	29	34
1970	25	30	34
1975	26	30	35
1980	29	32	40
1996	36	38	47

Table 4.2B Change in proportion of white-collar workers in the manufacturing industry, the engineering industry, and the chemical industry (average annual change in percentage points)

	Manufacturing	**Engineering**	**Chemicals**
1920–35	0.07	0.07	0.27
1935–60	0.52	0.52	0.60
1960–75	0.13	0.07	0.07
1975–96	0.48	0.38	0.57

Sources: SOS Industri 1930–1990 (Official Statistics of Sweden, Mining and Manufacturing).

manufacturing during the first transformation phase occurred both as a result of the relative growth of the engineering sector and of employment-structure shifts within sectors. When we control for changes in the relative size of sectors, the increase in relative employment of non-production workers in total manufacturing between 1935 and 1960 is reduced to 0.4 per cent per annum. This means that change overall was less pronounced than in the engineering and chemical industries. This confirms, to some extent, our hypothesis that employment shifts were most pronounced in leading sectors.

In order to test this hypothesis further, we computed employment shares for those subsectors that have been defined as 'leading' in different periods. For the structural cycle 1935–75, we chose three subsectors of the engineering industry: production of machinery, production of transport equipment, and production of electro-technical products. For the cycle beginning in the 1970s, we selected the pharmaceutical industry and electro-technical engineering, as the latter includes the ICT industry. The results are reported in Table 4.3. During the 1935–60 period, the level of change in the selected subsectors was essentially the same as that displayed at main sector

Table 4.3A Proportion of white-collar workers in selected industrial subsectors (percentage of the total work force)

Year	Machinery*	Transport equipment	Electro-technical engineering†	ASEA	Pharmaceuticals
1920	N/a	N/a	21	26‡	N/a
1935	17	N/a	23	27	N/a
1950	24	18	29	33	47
1960	30	25	35	43	55
1970	33	30	35	40	55
1975	33	29	36	40	59
1980	34	29	44	44	63
1990	40	30	47	N/a	69
1996	38	32	54	N/a	73

N/a = not available
* Excluding computers and office machines
† Including computers and office machines and optical instruments
‡ 1925

Sources: SOS Industri 1930–1990 (Official Statistics of Sweden, Mining and Manufacturing); Glete (1983).

Table 4.3B Change in proportion of white-collar workers in selected industrial sub-sectors (average annual change in percentage points)

Period	Machinery*	Transport equipment	Electro-technical engineering[†]	ASEA	Pharmaceuticals
1920–35	N/a	N/a	0.13	0.07[‡]	N/a
1935–60	0.52	N/a	0.48	0.64	N/a
1950–60	0.60	0.70	0.60	1.00	0.80
1960–75	0.20	0.27	0.07	−0.20	0.27
1975–80	0.20	0	1.60	0.80	0.80
1980–96	0.25	0.19	0.63	N/a	0.63

N/a = not available

* Excluding computers and office machines

[†] Including computers and office machines and optical instruments

[‡] 1925–35

Sources: SOS Industri 1930–1990 (Official Statistics of Sweden, Mining and Manufacturing); Glete (1983).

level in the engineering industry (that is, slightly more pronounced than in total manufacturing). For the rationalization phase between 1960 and 1975, the picture is mixed. Developments in machinery and electro-technical equipment paralleled those in total engineering. In transport equipment and pharmaceuticals, the upgrading of skills continued to be above the industrial average, contrary to expectations. This may be explained by the fact that both industries introduced new technologies during the period, with the result that investments were not directed towards rationalization measures. In the case of transport equipment, the anomaly is due to the growth of a new activity, aircraft production, which also carried new technologies. When this subsector is excluded, the increase in the share of non-production workers in the transport equipment industry disappears. As expected, the rise in relative employment stagnated in the mature parts of the subsector, while stagnation for the subsector as a whole was postponed for a decade.

The structural cycle that was based on heavy engineering and heavy chemicals came to an end with the crisis of the mid-1970s. In the transformation of Swedish industry that followed, electronics and biotechnology stood out as the leading sectors. In our data, the former is included in the electro-technical industry, while the latter is represented by the pharmaceutical industry.

As shown above, relative employment of non-production workers accelerated after 1975, both in the engineering and chemical industries, and in manufacturing at large. It is clear, however, that this tendency was more pronounced and also occurred at an earlier stage in leading subsectors. The pharmaceutical industry stood out as different from the more mature parts of the chemical industry; it would help to reshape the economy in later decades by exploiting new innovations in biotechnology. The development potential for these activities was in fact established in the 1970s.[5] A key factor was a 70 per cent increase in the number of non-production workers between 1970 and 1980, raising their proportion of the total work force in the subsector from 55 per cent to 63 per cent. As a result, expected employment structure shifts came earlier, and were more pronounced, than in most other industries.

A similar pattern can be observed in the electro-technical industry. Although the share of non-production workers in most parts of the engineering industry did not rise until 1980, this increase had begun in the electro-technical subsector in 1975, and at a substantially faster pace.

The direct evidence of labour-demand shifts in the engineering industry that is available for the 1960–90 period further supports our hypotheses. Table 4.4 shows that the reported relative shortage of technical staff was at a significantly lower level between 1960 and 1975 than between 1976 and 1990. As expected, this pattern was more pronounced in the electro-technical industry than in mechanical engineering.

Table 4.4 Ratio of firms that reported a shortage of technical staff to firms that reported a shortage of semi-skilled and unskilled workers (averages for the periods 1960–75 and 1976–90)

Period	Total engineering	Mechanical engineering	Electro-technical
1960–75	1.3	1.2	2.9
1976–90	6.4	4.9	9.2

Source: My computations from *Konjunkturbarometern (Business Cycle Barometer)*, March, Konjunkturinstitutet (National Institute of Economic Research) 1960–90.

Institutional change

Wage policies are a type of institution that, if turned into wage agreements, have a significant influence on wage structure. A wage policy rests on a particular ideological foundation. During our period, the idea of 'solidarity' was a widely-articulated concept. In the trade union movement it was known as 'solidaristic wage policy', and was associated with the Rehn–Meidner model and the process of wage equalization of the 1960s and 1970s (Lundberg 1985). Prior to that, it had a long history in both the ideological and practical debates within the Swedish Confederation of Labour (LO). An ideology of wage equalization had been articulated at LO congresses during the 1920s (Ullenhag 1971, p. 32), with ideas introduced by workers in sectors that were exposed to foreign competition, notably the metal and engineering industries. The unbalanced development of Swedish industry during the turbulent years after the First World War had created a substantial wage premium in favour of workers in sheltered sectors. The ratio of wages in sheltered sectors to those in exposed sectors rose from 0.95 to 1.25 between 1916 and 1922. The gap was not closed until the late 1930s (Svensson 1995, p. 81).

Construction workers and workers in the food and beverage industry were among the least affected by the fall in nominal wages during the deflation in 1920–22. Although blue-collar wages fell on average by 32 per cent, the decrease in these two sheltered sectors was only half of that (Svensson 1995, p. 75). The obvious reason was that employers could be compensated for wage costs by raising prices, since the lion's share of production consisted of non-traded goods with low price elasticities.

In a motion to the 1922 LO congress, the Stockholm section of the Swedish Metal Workers' Union declared: 'To the extent that the wage increases of other groups of workers exceed those of the workers in the export industry, the wages of the latter will be further reduced, something which can hardly be in accordance with the demands of class solidarity' (Svenska Metallindustriarbetarförbundet, 1922). The argument was that price increases on products from sheltered industries, which were often 'necessary also for the working class', directly reduced the real wage.

The metal workers repeated their claims for solidaristic wages at the 1926 LO congress (Ullenhag 1971, pp. 28–9). The arguments were the same as at the earlier congress, but this time their call for a wage policy based on 'class solidarity' was combined with proposals for centralization of power within the organization. These arguments were repeated throughout the 1930s.

Solidarity among employers

Solidarity was not confined to the union side, however. As demonstrated by Swenson (2002), solidarity among employers was a featured idea within employers' organizations, notably in the metal trades, even before the First World War (Swenson 2002, pp. 78 ff.). It implied a common strategy in wage-setting with the explicit purpose of preventing a cost-driving competition for labour. The link between wage policy and the pressure of competition is obvious. Since the drive to cost-reducing measures as a result of intensified competition tended to happen during phases of rationalization, we would expect employers to push hard for solidarity during these periods. In broad terms, this would be the period of the 1920s, along with the 1960s and early 1970s. We would also expect employers in sectors exposed to foreign competition to be more active in this process.

At least from the early 1920s, the authority of the Swedish Employers' Confederation (Svenska Arbetsgivareföreningen, or SAF) was used as a means of controlling the wage policy of the confederate organizations (Faxén 1991, p. 70). This may have been an effect of the creation of a united employers' organization in 1917 when the Swedish Engineering Employers' Association (VF) joined SAF. By virtue of its relative size, VF soon became the most influential of the member organizations (Törnqvist 1954).

The problem of keeping wage growth in the sheltered sector, particularly the building trade, in line with development in other industries had attracted attention before the war. In his 1924 annual report the Chairman, von Sydow, complained that SAF had not managed to solve the problem of employers in the sheltered sector, 'that practised the law of least resistance instead of showing solidarity with their fellow employers in the exposed sector' (Faxén 1991, pp. 73–4).

The drive to centralization

Thus within LO, as well as within SAF, representatives of the metal trades argued for what were labelled 'solidaristic wages'. These efforts had limited results, partly because of unfavourable organizational structures and power relations. During the 1930s, efforts were devoted to the restructuring of the organizations of employers as well as workers. For SAF the process was completed in the mid-1930s, when employers in the engineering industry established considerable power within the organization.[6]

Beginning in 1933, a major conflict in the building trade brought the problem of wage differentials between sectors to a head. It prompted employers as well as workers in the exposed sector to fight internal battles against their sheltered sector fellow members. By the end of the 1930s, power had been centralized in the hands of exposed sector agents, notably from the engineering industry, within both SAF and LO (Swenson 1991, pp. 521 f.).

When the institutional setting was altered in favour of the implementation of a solidaristic wage policy, the demand for wage equalization had declined. The economy was already in a phase of transformation and the structure of labour demand had shifted in favour of labour with higher skills. The fierce competition in all markets that is characteristic of the rationalization phase was replaced by less strict conditions for firms that applied new technologies and modes of production. This new situation both demanded and provided scope for labour-attracting relative wage increases in leading sectors. The ratio of male wages in the engineering industry to male wages in the food and beverage industry increased from 0.89 in 1932 to 1.05 in 1942.

The implementation of solidaristic wages

In 1951 wage equalization became an integral part of a new programme for the promotion of economic growth with both full employment and price stability: the Rehn – Meidner model. However, the first agreements between LO and SAF that featured provisions designed to benefit low-paid workers were not signed until 1964. Between 1969 and 1974 the equalizing tendency in the wage agreements became even more pronounced (LO 1986). Considering the long history of the solidaristic wage doctrine, it is reasonable to ask why this policy was not implemented until the 1960s. The traditional view holds that it was not until then that the LO had gained the power to force through wage equalization, although such a policy had constituted a fundamental part of trade union ideology for some time (Edin and Holmlund 1995).

An alternative *cross-class alliance* approach (Swenson 2002) would focus on differences in conditions and interests between sectors rather than between labour and capital. It may well be that the struggle by the metal workers for fair wages arose out of self-interest, before becoming a constituent part of trade union ideology. Equally important were the fierce competition and cost-reducing prospects of rationalization that made employers, particularly in the engineering industry, favourably disposed to wage equalization. There are clear indications that employers in the engineering industry approved of wage equalization. Collective agreements between the Swedish Metal Workers' Union and the Swedish Engineering Employers' Association already had a 'low-wage profile' from 1961 (Svenska Metallindustriarbetarförbundet 1962, pp. 23 f.). This means that wage compression in this sector had begun before the first 'solidaristic' central agreements between LO and SAF had been signed.[7]

The solidaristic wage policy became an efficient means of attracting new groups, notably women, to the labour force. Employers, particularly those in expanding firms in the engineering industry, sought ways of filling job vacancies with low initial skill requirements by tapping into the labour force reserve of married women.[8]

In 1960 LO and SAF agreed to abolish the separate wage tariffs for male and female workers over a period of five years beginning in 1962. The implementation of the policy varied among sectors, both with respect to scope and pace. It was most extensive and most rapid in the engineering industry, which indicates that the growth of demand for female labour was related to the rationalization process in manufacturing. The highly Fordist organization of production in parts of the engineering industry (that is, the automotive industry) made it profitable to employ more women in the factories. Productivity crests were short, which rendered negligible the drawback of higher expected female turnover compared with male turnover. The advantage of lower female wages was more important. Job vacancies with low initial skill requirements could be filled at relatively low costs by drawing upon the labour reserve of married women.

The evidence indicates that wage increases were a successful means of attracting women to factory work; and in male-dominated workplaces it was vitally important to supplement the work force without having to increase male wages. Since such a strategy would include raising female relative wages, the proposal to abolish separate wage rates for men and women ought to have furnished a golden opportunity to effect the desired change in the wage structure.

The automotive industry provides an example.[9] While the female-to-male relative wage in this industry increased from 73 per cent in 1960 to 96 per cent in 1965, the male wage as a percentage of the average male wage in total manufacturing decreased from 122 per cent to 105 per cent during the same period. Between 1960 and 1975, the number of female workers increased by no less than 750 per cent and the proportion of female workers rose from 2 per cent to 15 per cent of the total work force.

The dismantling of the Swedish model

In the mid-1970s, the two-part centralized bargaining model started to erode. One reason was the growth of services, particularly public services, which shifted the balance of power in the labour market. Another reason that has less frequently been emphasized was related to the structural crisis of the 1970s, and the subsequent transition to a phase of transformation and renewal of Swedish industry based on innovations in the areas of electronics and biotechnology.

This brought a shift in the primary function of the wage structure from cost reduction, which had dominated during the previous rationalization phase, to labour allocation. Wage differentials became an important means of attracting and keeping workers, who possessed the skills required to exploit the productivity potentials of the new technology. Just as in the 1960s when the engineering industry, squeezed by fierce international competition, took the lead in applying rationalization measures to reduce production costs, leading firms in the same sector pioneered the application of the new technology of the electronic era. This may explain why the Swedish Engineering Employers' Association (Verkstadsföreningen, or VF) came to question the value of the central agreements in the early 1980s, in particular the low-wage provisions and the 'wage development guarantees' that compensated groups with small or non-existent wage drift.

Opinion within VF favoured the abandonment of central bargaining, and it managed to conclude a separate agreement with the Swedish Metal Workers' Union in 1983. Policy formation later in the decade revealed that workers in the engineering industry also saw the need for change. At the 1989 congress of the Swedish Metal Workers' Union (Metall), a document entitled 'Solidaristic work policy' laid the groundwork for a shift from equalization of wages to a wage policy that emphasized the importance of rewarding competence and skills. At the rhetorical level, this shift represented a continuation of the arguments for solidarity as an engine of growth featured in the Rehn – Meidner report of 1951. However, this called for a redefinition of *solidarity*. The objective of the solidaristic work policy was to transform all work to 'good work'. The solidarity component would be satisfied by providing the prospect of developing the workers' competence and skills rather than by wage equalization. This would also contribute to productivity growth, and thereby enhance the ability to increase real wages (Svenska Metallindustriarbetarförbundet 1989, pp. 7 ff.).

Industrial dynamics, labour market institutions, and the role of leading sector agents

The history of the solidaristic wage policy may now be interpreted within our structural-economic framework. It was first formulated in the early 1920s, as fierce international competition limited wages in the metal trades and other exposed sectors. Wage equalization was not only in the interest of workers in the metal trades but also of employers in this exposed sector, because of their inherent disadvantage in the competition for labour between sectors. Existing labour market institutions did not support a solidaristic wage policy, however. During the 1930s, power within both LO and SAF became centralized in the hands of workers and employers in the engineering industry, and within the framework of the Rehn – Meidner model the bargaining system was centralized and co-ordinated. Thus, when competition again grew fierce, towards the end of the 1950s, the solidaristic wage policy could be implemented and, in combination with Fordistic rationalization, serve as a means to force unit costs down. As the potential of rationalization reached its limits, the structural crisis in the mid-1970s marked the transition to a phase of transformation based on electronics and biotechnology. Complementarity between new technology and skills altered the structure of labour demand, and wage equalization came to an end. Leading sector agents worked to break up the coordinated and centralized bargaining system, and solidarity was reformulated to motivate wage differentials on the basis of skill differentials. This development clearly supports our hypotheses that institutions that promote wage convergence tend to be implemented during rationalization, and institutions that promote wage divergence tend to be implemented during transformation; that this tendency is most pronounced in leading sectors; and that leading sector agents are the most active promoters of institutional change.

Final remarks

Two sets of observations lend strong support to the hypothesis that labour market institutions have contributed to the capacity of leading sectors in Swedish industry to absorb and exploit new technologies. First, changes in wage and employment structures have been closely related to systematic differences in the direction of technological change. Periods of rationalization were connected with strong equalizing tendencies, while equalization-retarded and relative employment of non-production workers rapidly increased during periods of technological and structural transformation. The difference mirrors a shift in the centre of gravity from a cost-reducing function to a labour-allocating function of the wage structure. The tendencies were most pronounced in leading sectors.

Second, the development of informal and formal labour market institutions followed a path that conformed tightly to the observed pattern of change in wages and employment. Two observations lead us to believe that renewal of labour market institutions was intended to fulfil the changing demands of firms in leading sectors. First, shifts in employment and wage structures, which indicate changes in capital – skill relations, preceded institutional change. Second, leading sector agents, both workers and employers, were often the initiators of institutional change, sometimes in direct conflict with laggards.

Notes

1 The financial assistance of the Swedish Council for Work Life Research (grant no. 2000–0286) is gratefully acknowledged.
2 Complementarity in this context is understood as an increase in the marginal productivity of one factor (capital or labour) resulting from the input of the other factor (capital or labour).
3 Theoretically, these changes in wage ratios may originate from shifts in supply as well as shifts in demand. A supply-side explanation is discussed and refuted in Svensson (2003).
4 ASEA (ABB after a merger with Swiss Brown Boveri in 1988) is a technically-advanced producer of equipment for power generation, power transmission and communication, as well as a pioneer in the production of industrial robots.

5 See Dahmén (1997), who stresses the importance of 'embryonic development blocks', which originated in the 1920s, for the transformation of the Swedish economy after the Great Depression.

6 Sigfrid Edström, managing director of ASEA, one of the leading firms in the engineering industry, and chairman of VF since 1916, was elected chairman of SAF in 1931. He was a dedicated centralist and would play a major role in the future development of the organization. Another key player was Gustaf Söderlund, who was appointed managing director of SAF also in 1931 (Treslow 1986, pp. 19 ff.).

7 This interpretation is further supported by the continuous decrease in wage dispersion among blue-collar workers in the engineering industry from 1961 until 1980, as reported in Svensson (1995), Figure 8.2.

8 A conference entitled 'Women as a labour resource', and arranged jointly by the LO and SAF in Saltsjöbaden outside Stockholm in 1964, serves as an example of these efforts. Four out of five representatives of individual firms were from large firms in the engineering industry (Arbetsmarknadens Yrkesråd 1965, p. 34).

9 This was one of the most expansive areas of the engineering industry with an increase in share of total exports from 1.5 per cent in the early 1950s to 6 per cent in 1960 and 11 per cent in 1970.

References

Abramovitz, M. (1986), 'Catching Up, Forging Ahead, and Falling Behind', *Journal of Economic History*, 46, pp. 385–406.

Arbetsmarknadens Yrkesråd (1965), *Kvinnorna som arbetskraftsresurs*. En konferens i Saltsjöbaden anordnad av Arbetsmarknadens Yrkesråd (Stockholm: Arbetsmarknadens Yrkesråd, SAF and LO).

Berman, E., J. Bound and S. Machin (1998), 'Implications of Skilled Biased Technological Change: International Evidence', *Quarterly Journal of Economics*, 113, pp. 1245–80.

Dahmén, E. (1950), *Svensk industriell företagarverksamhet* (Stockholm: IUI).

Dahmén, E. (1988), 'Development Blocks in Industrial Economics', *Scandinavian Economic History Review*, 1.

Dahmén, E. (1997), 'The inter-war years: industry in transformation', in L. Jonung and R. Ohlsson, *Economic Development of Sweden since 1870* (Lyme: Edward Elgar).

Edin, P. A. and B. Holmlund (1995), 'The Swedish Wage Structure: The Rise and Fall of Solidarity Wage Policy?', in R. Freeman and L. Katz (eds), *Differences and Changes in Wage Structures* (Chicago, IL: University of Chicago Press).

Faxén, K.-O. (1991), 'Några kommentarer till SAF:s lönepolitiska uttalanden under 1920 – talet', in *Vingarnas trygghet. Vänbok till Gösta Rhen* (Lund: Dialogos).

Freeman, C. and F. Louca (2001), *As Time Goes By: From the Industrial Revolution to the Information Revolution* (Oxford: Oxford University Press).

Goldin, C. and L. F. Katz (1998), 'The origins of technology–skill complementarity', *Quarterly Journal of Economics*, 113, pp. 693–732.

Konjunkturbarometern (Business-Cycle Barometer), 1960–1990 (Stockholm: Konjunkturinstitutet).

Ljungberg, J. (1990), *Priser och marknadskrafter i Sverige 1885–1969. En prishistorisk studie* (Lund: Skrifter utgivna av Ekonomisk-historiska föreningen, Vol. LXIV).

LO (1986), *De centrala överenskommelserna mellan LO och SAF 1952–1987* (Stockholm: LO).

Lönestatistisk Årsbok (Stockholm: Socialstyrelsen), 1928–1951.

Lundberg, E. (1985), 'The Rise and the Fall of the Swedish Model', *Journal of Economic Literature*, XXIII (March), pp. 1–36.

North, D. (1990), *Institutions, Institutional Change and Economic Performance* (Cambridge: Cambridge University Press).

Perez, C. (2002), *Technological Revolutions and Financial Capital* (Cheltenham: Edward Elgar).

Schön, L. (1998), 'Industrial Crises in a Model of Long Cycles: Sweden in an International Perspective', in T. Myllyntaus (ed.), *Crisis and Restructuring in Small Countries* (Berlin: Scripta Meracaturae).

Schön, L. (2000), *En modern svensk ekonomisk historia. Tillväxt och omvandling under två sekel* (Stockholm: SNS förlag).

Schön, L. (2003), 'Swedish Industrialization and the Heckscher – Ohlin Theory', Paper for Eli Heckscher 1879–1952: A Celebratory Symposium at Stockholm School of Economics, 22–24 May 2003.

SOS Industri (Stockholm: Kommerskollegium 1930–60; Statistiska Centralbyrån 1961–90).

SOS Löner (Stockholm: Socialstyrelsen 1952–60; Stockholm: Statistiska Centralbyrån 1961–95).

SOU (1976), *Arbetstidsförkortning – när? hur?*: 34 (Stockholm: Liber Allmänna Förlaget).

Svenska Metallindustriarbetarförbundet (1922), Metall, avdelning 1, Kongressmotion, Arbetarrörelsens arkiv och bibliotek, Stockholm.

Svenska Metallindustriarbetarförbundet (1962), *Verksamhetsberättelse 1962* (Stockholm: Tiden).

Svenska Metallindustriarbetarförbundet (1989), *Solidarisk arbetspolitik för det goda arbetet* (Stockholm: Metall).

Svensson, L. (1995), 'Closing the Gender Gap. Determinants of Changes in the Female-to-Male Wage Ratio in Swedish Manufacturing 1913–1990' PhD Dissertation, Ekonomisk-historiska föreningen vid Lund University, Lund.

Svensson, L. (2003), 'Industrial Dynamics, Labour Demand, and Wage Differentials in the Swedish Manufacturing Industry, 1930–1990', in C. Lundh, J. Olofsson, L. Schön and L. Svensson (eds), *Wage Formation, Labour Market Institutions and Economic Transformation in Sweden 1860–2000* (Lund: Lund Studies in Economic History), Vol. 30.

Swenson, P. (1991), 'Bringing Capital Back in, or Social Democracy Reconsidered', *World Politics*, 43 (July), pp. 513–44.

Swenson, P. (2002), *Capitalists against Markets* (New York: Oxford University Press).

Tegle, S. (1982), 'Den ordinarie veckoarbetstiden i Sverige1860–1980', unpublished manuscript, Department of Economics, Lund University.

Treslow, K. (1986), *Verkstadsföreningen under 90 år* (Stockholm: VF).

Törnqvist, E. (1954), 'Verkstadsindustrin, ett maktcentrum i svensk avtalspolitik', *Statsvetenskaplig tidskrift*, 3.

Ullenhag, J. (1971), *Den solidariska lönepolitiken i Sverige. Debatt och verklighet* (Stockholm: Scandinavian University Books).

5
Did Higher Technical Education Pay?[1]

Jonas Ljungberg

The issue

Despite the general agreement about the importance for economic growth of higher education, and in particular higher technical education, facts that prove or disprove this notion are not easily available. One approach would be to disaggregate national accounts and look at sectoral data for education. However, such data are seldom reported in the official statistics. Moreover, in national accounts, education, like other non-marketed services, suffers from the non-existence of price and output estimates. Labour input – that is, teachers and other personnel – is therefore taken as a proxy for the contribution of education to GDP. In the most recent system for national accounts (SNA 1993, pp. 402–3) other measures are also discussed, but for historical periods national accounts rely solely on input data. For the study of change over time that poses a problem, since current values must be deflated into constant prices. And the labour input is deflated with an index of the relevant earnings which means that the estimated 'output' actually equals the quantity of inputs. In other words, nothing is, by definition, left for productivity change. The contribution of the education sector to economic growth thus equals its increase of employment, at the value of labour in the chosen base year. The sources of the part of growth contained in the 'residual', that is the total factor productivity, should thus be sought somewhere other than in education. Or should we look for new approaches to this fundamental problem?

The present chapter intends to shed some light on the formation of human capital in Swedish higher technical education since about 1870, as part of a more general intention to discuss the problem of measurement of human capital. It is thus not about the role of human capital in economic growth but rather about how to provide the data for an analysis of that problem. How can education be measured, here focused on higher technical education and with quantitative aspects of its long-term development? Higher technical education considers primarily the formation of engineers and architects.

As is shown by McInnis in Chapter 3, formal education was not the only way to become an engineer in late nineteenth-century Canada, and the same was the case in Sweden (Torstendahl 1975, p. 46). Besides self-made learning-by-doing in production, the military provided engineering knowledge also for the civil economy. Military engineers filled the salary lists of the railway building projects at least until the 1880s, with titles such as lieutenant, captain and major or colonel relating to the works hierarchy.[2] Ahlström (1982, p. 40) estimates that in 1850 the Swedish stock of engineers with higher technical education numbered 1,100–1,200 individuals, of which a good half were produced by the military and about a quarter each by the Royal Technical Institute in Stockholm, and the Chalmers Institute in Gothenburg. From the 1870s the education at these institutes evolved on a larger scale and in the 1890s about 100 engineers graduated each year. In the decades after the second World War several new technological institutes were established at the general universities and higher technical education expanded greatly.

The chapter is organized as follows. The next section gives an account of the expenditures of the technological institutes. Contained in the expenditures is also a conventional national accounts estimate of higher technical education. This is followed by a section which changes the focus to the micro level and explores the long term development of relative earnings of engineers. The premium on higher technical education is calculated in the section after that. It is based on the difference in earnings between graduate engineers and college engineers. The former were those trained 4 years (from 1986 a half year was added) at the university level while the college engineer got an exam from technical institutes at the secondary level. In the literature estimates of the return to education are usually based on cross-sections of more recent date, by necessity therefore also including a sizeable element of crystal-ball gazing concerning the development of earnings. The present study contributes with an estimate of the premium on higher technical education in Sweden back to 1871. Moreover, inspired by an approach tried by Jorgenson and Fraumeni (1992) for the whole education sector of the USA over 1948–86, these premiums are aggregated to an output estimate for the higher technical education in Sweden over 1871–1992. Thus we have returned to the macro level of the economy and this output estimate is compared with a conventional national accounts estimate. The final section raises a question about the measurement of human capital and considers whether higher technical education did pay, for the individual and for the society.

Expenditures on higher technical education

Broadly speaking Swedish higher technical education developed its present structure in the late nineteenth century. The government increased its support to the Royal Technical Institute, changing its name to University (actually *Högskola*, after the German *Hochschule*) in 1876, and to the

Chalmers Institute. The latter was originally run by a private foundation but also received support from the city of Gothenburg and the government. Both schools were, however, in the realms of the parliamentary solicitors and were included in their annual report on public services. These reports are the source for the construction of series on labour cost, investments and intermediate costs, summing up to the expenditure series shown in Figure 5.1. Total expenditures were reported and it is not always possible to distinguish how much was provided by government and how much raised from other funds. In the 1930s the accounts were omitted in the solicitors' reports, and instead government accounts have been used as a source up to 1992.[3] In the latter only government funding of the expenditures is reported and this shows up in Figure 5.1 as a break, with a new expenditures curve starting from a lower level in the inter-war years. With the expansion from the 1940s the role of private funding diminished, but on the other hand separate public funding of research became more and more important from the 1960s. In addition, private foundations took part in this 'external funding', so that researchers have to apply and compete for financial support. From the mid-1960s to the early 1990s the public expenditure for 'external research' increased more than seven times, in real terms.[4] However, even if this research has an important feedback on higher education, it is not an immediate part of it and its funding has not been included in the expenditures here. The private funding before 1930 could (to simplify somewhat) be assumed to have been allocated for research, meaning that only government

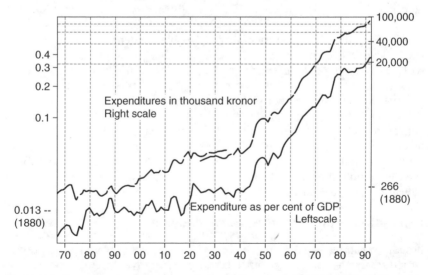

Figure 5.1   Expenditures on higher technical education and GDP, 1867–1992
Sources: See text and for details Ljungberg (2004b); GDP from SHNA.

expenditures are included in the calculation of the share in GDP of the expenditures on higher technical education.[5]

This notwithstanding, there *is* some research that cannot be excluded, namely the 'internal' research connected with professors' positions and so forth. For most of the period there are no clues in the data as to how this part could be separated, and neither is there any sound reason to do so. The 'internal' research performed within the ordinary university positions, in the Swedish system, must be seen as an institution for the feedback so necessary between research and higher education. Even if the research financed by external funding also has a feedback on the higher education, it is more independent and its proportions vary widely between the different institutes. Therefore it is reasonable to include the 'internal' but to exclude the 'external' research in the expenditures of higher technical education.

A few things stand out from Figure 5.1: first, the steady increase in the expenditures on higher technical education, shown here in constant prices of 1910/12;[6] second, their acceleration in the early 1940s; third, some periods with slower growth or moderation: before 1890, the inter-war period, and after 1980, at least compared with the preceding four decades. All higher education received more resources in the 1940s but higher technical education began the acceleration a couple of years before the other sectors, in the middle of the war. However, from the end of the war until the early 1970s, higher technical education actually lagged behind and its share of the expenditure on higher education almost halved. A new turnaround occurred in the early 1970s and in 1992 its share in higher education was back at the level of 1945.[7] If higher technical education has been a driving force for economic growth, whereas other higher education has not, then the time pattern is complex and with a gestation period.[8] Fourth, compared with GDP, expenditures have increased broadly along a similar pattern, with moderate increase up to 1940 and a more rapid increase thereafter. Some differences, in particular before 1940, can nevertheless be discerned. Thus the long term increase compared with GDP came in short spurts, before 1880, 1920 and 1948, as well as during the long post-war increase. These 'spurts', and also the jump before 1980, reflected more a slowing down in GDP growth than an increase in expenditures on higher technical education, with the notable exception of the spurt in the 1940s which came along with a remarkable growth in GDP above 6 per cent a year (1943–49). After 1940 higher technical education has thus expanded more than GDP, both initially and in the long term.

With the shift to rapid expansion of higher technical education, the composition (in current prices) of expenditures also changed. Investment, which besides buildings include durable equipment, apparatus and books, more than doubled its share from 11 per cent in the 1930s to 27 per cent in the 1940s, and then stayed around a quarter of total expenditures. Contrary to what might be expected, the share of labour cost has declined in the very long term, being above two-thirds for the period before 1914 and roughly half for the whole period 1945–92. After peaking at 65 per cent in the early 1960s the share has been declining, to 50 per cent in 1980 and 47 per cent in 1992. The capital, included as investments in the total expenditure here, is an unresolved issue but so far has not been included in the government sector of national accounts (Levin 1996, p. 200). As a consequence, the share of labour cost in total expenditure just reported is smaller than the share of value added in national accounts. Broadly speaking, both shares have developed along similar lines, the share of value added averaging 77 per cent before 1914 and oscillating closely around 65 per cent after 1970. To sum up about the composition of the expenditures of higher technical education: labour costs, or value added in national accounting, have developed in a pattern similar to that of total expenditure in Figure 5.1, only with minor deviations and growing slightly slower.

Relative earnings since 1867

Have the expenditures on higher technical education, representing invest-ments according to economics of education, returned any benefits? Whether we approach this question from the point of view of the individual or the society should not be so different, at least when treated as an accounting exercise. For the individual the benefit is not immediately dependent on the size of the expenditures, but on whether the extra years in school result in a sufficient increase in lifetime earnings. The investment made by the indi-vidual, or by the family of the individual, calculated as the extra costs and forgone income when in school, should pay off in higher earnings that could be recalculated as an 'internal rate of return'. The internal rate of return is the interest rate at which the cost of investment equals the discounted value of the higher earnings. In actual fact the internal rent can only be calculated *ex post* but usually the future incomes, or part of them, are extrapolated. For education to be profitable, the level of the internal rent should exceed other investment opportunities.

I perform a less ambitious task and calculate only the premium on edu-cation: that is, the extra income over the lifetime less the cost, discounted in the year of graduation from school. The development of the premium on education over time gives a view of how society – both markets and institutions – has valued higher technical education.

A starting point for the exercise gives Figure 5.2, showing the earnings differential between a graduate engineer and the average male worker in manufacturing, as well as between a graduate engineer and an college engin-eer, since early industrialization in Sweden. Before commenting the pattern the data must be spelled out a little. Clearly defined earnings for different grades of engineers are available first from statistics of the SAF dating from 1952. Unfortunately these statistics have remained confidential and only

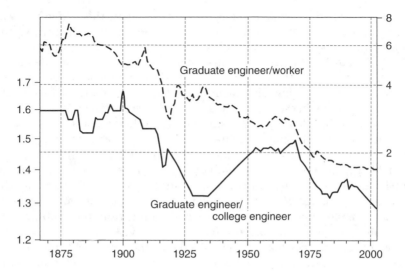

Figure 5.2   Relative earnings of a graduate engineer over a college engineer and a worker in manufacturing, 1867–2002

Sources: See text and for details Ljungberg (2004b).

the period 1952–86 has so far been covered accurately. After 1986 the official statistics have been used. Up to 1934 the earnings of graduate engineers are based on the records of those employed by the State Railways (Statens Järnvägar), combined with the relevant salary scales for government employees. Since changes in the salary scales were not frequent, it has been enough to investigate benchmark years and to reconstruct annual variations from other sources. In the years 1917–24, when earnings were affected first by inflation and then by deflation, engineers at the Kockums shipyard in Malmö, across the Sound from Copenhagen, have been used for a cross-check. The same cross-check could be performed for 1912; and for the first two decades or so, 1867–85, the remuneration of engineers at the state railway buildings has been compared with those employed in the operation of the railway traffic (Records of Kockums; Records of Statens järnvägsbyggnader). Those in service at the railway buildings at first sight seem to have been paid substantially more, but when taking account of the housing compensation adding 15–20 per cent to the salary scale of State Railways employees, the difference is marginal.[9] There was high status attached to working with the railways, rather like IT professionals today, and employees were paid accordingly (*Tullkommitténs* 1882, p. 266). In 1912 the railway engineers, we may assume from the shipbuilder records, may well have earned about 10 per cent more than their colleagues in private industry.

On the other hand, the chance of advancing to leading positions with much higher remuneration was greater in industry. However, with the growing numbers of graduate engineers in industry this difference was reversed, and the earnings of the railway engineers confirm the saying, 'the cake of the state is small but secure'. After 1934, the railway engineers lagged behind and in 1952, there was a huge gap (49 per cent) to graduate engineers in industry, as revealed by the SAF statistics available from this year. There are at least two possible explanations. One is that more engineers who graduated from higher secondary school colleges were employed by the State Railways and thus depressed the earnings. The other is that the fast rise of earnings in private business, during the war and the early post-war period, was not matched by those in public service. In 1956 there was still a considerable gap between privately employed graduate engineers and those in public service, the former in their forties earning 35 per cent more than the latter of the same age. Since the vast majority of the graduate engineers were privately employed an earnings series based on the, from the 1940s, poorly paid railway engineers would not be representative.[10] From 1934 to 1952 the earnings series for graduate engineers has therefore been interpolated according to the trend and with annual variations as in the earnings series for 'technicians'.

The earnings series for 'technicians' in the official statistics runs from 1913 and in 1952 exactly matches the SAF statistics for privately employed college engineers. I have therefore let the official technician series denote college engineers. In 1914 the level is slightly adjusted with the 'lower engineers' in the State Railways record, a position introduced around the turn of the century. For the nineteenth century the earnings data for this group are more sparse, and the series is partly extrapolated from the series for graduate engineers.

Despite the mutual interdependence of the two series on earnings for different categories of engineers, Figure 5.2 reveals a changing relation over time. Over the long term the relative earnings of graduate engineers have declined much less compared with college engineers than when compared with workers, but nevertheless a clear decline. Particularly striking is the considerable decline after 1970 that was only briefly interrupted in the 1980s. The continuation of the decline after 1990, which is contrary to international observations of income distribution, might be explained by a substitution of engineers with university diplomas or a bachelor of science degree, for college engineers. In 1989 technical universities began also to offer a university diploma in engineering with two years of study, and in 1993 a bachelor of science in engineering with three years of study. The engineering degree accordingly disappeared from the realms of secondary school colleges in the early 1990s.

To settle whether the decline after 1990 in relative earnings of graduate engineers is a fact or an artefact one needs more detailed statistics than those used here. The reader who has spotted the Swedish periods of 'transformation' and 'rationalization', in Chapters 1 and 4, may find a connection here.

Lars Svensson's hypothesis in Chapter 4, about a higher demand for skilled labour during transformation periods and the reverse during rationalizations, gains some support from the relative earnings of graduate engineers over college engineers. The increases in their relative earnings in the 1890s, from the mid-1930s to the mid-1950s, and again in the 1980s, thus occur in periods of transformation. No such pattern can, however, easily be seen in the relative earnings over the average worker in manufacturing. However, according to the hypothesis there were also variations between the demand for skilled and for unskilled workers. If the demand for graduate engineers and for skilled workers covaried, then the periods with increased demand for knowledge and skills will not be discernible in the relative earnings of graduate engineers over the average of wages in manufacturing, since the average wage of workers will also increase when demand for skill is high. The long term decline in the relative earnings of a graduate engineer over the average worker is anyway striking, and it raises a question to which I will return in the concluding section: can human capital be defined only by the earnings differential, whereas earnings up to the level of a worker represent 'raw labour'?

The premium on higher technical education

For the present exercise I will, however, focus on the earnings differential between the graduate engineer and the college engineer. This differential is particularly relevant here, for two reasons. First, since the the two careers are very similar, it could be argued that the difference in their earnings throughout life, and hence in their human capital, is explained by the difference in their respective formal education, and not by experience. Superficially this presumption is contradicted by the steeper age profile of earnings for a graduate engineer compared with a college engineer, giving the impression that the human capital of the graduate engineer is substantially formed by experience on the job. However, the shape of the age profile could be seen as a function of educational training, with different functional forms for different education programmes.

The second reason for the particular relevance of the differential between graduate engineers and college engineers is that the alternative career for a graduate engineer, if not directed into a field of studies other than technology, would normally be to work as a college engineer. In the calculation of the premium on higher technical education, I have thus assumed that instead of four years at a technical university our subject would enter upon a career and earn an income as an engineer of lower rank. I assume, though, that the graduate engineer has worked for two months during the holidays each year, at a level of pay similar to a college engineer of the corresponding age, but gaining seniority more slowly. The forgone income during four years is accumulated with interest until the year the graduate engineer graduates. The difference in earnings between the graduate engineer and the college engineer during the following career is then calculated, and discounted backwards to the year of graduation. It is assumed that the graduate engineer graduates and enters his or her professional career at the age of 25.[11] The youngest cohort for which actual earnings can be used, for an estimate of the premium on higher technical education up to 2002, is that born in 1937 and entering the career in 1962. For those born after 1937 the premium is partly calculated on the earnings profiles of 2002, and extrapolated to 2042. The future will show whether the expected premium on higher technical education from 1963 onwards will turn out differently *ex post* from the present estimate.

Before looking at the estimated premium on higher technical education, allow me a few words on earnings age profiles. These are very important since the relative income in the early years after graduation has a greater impact and, the higher the age, the lesser the impact on the premium on education. Figure 5.3 shows that the relative earnings of 25 year old graduate engineeers and college engineers, compared with older colleagues, have changed significantly over time. For the period 1952–77 the age profiles could be based on actual data for the entire population of engineers employed in SAF enterprises, and the same goes for 1986. Thereafter only a crude age profile in the official statistics for 2002 could be used for an estimation. Before 1951, age profiles have been estimated from the engineers employed by the State Railways. The SAF statistics show that the age profiles in the 1950s were much flatter for engineers in public service than in private business. One could therefore presume that the high relative earnings for young engineers before 1940 are due to a bias in the State Railways material. However, a change

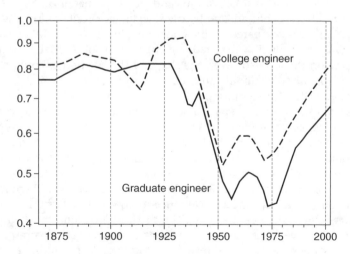

Figure 5.3   Salaries of 25 year old relative to 40–45 year old engineers, 1867–2002

Sources: See text and for details Ljungberg (2004b).

occurred during the 1940s in the age profiles of the railway engineers, and also the data on Kockums engineers in the early twentieth century indicate a flatter age profile. Since few railway engineers were under the age of 30, estimations of complete age profiles have been constructed by fitting equations to the data. Hopefully doing this adjusts for most of the bias in the data.

It is interesting to note that the relative earnings of young engineers were at a low during the 'Golden Age' of economic growth. The age profile of earnings was steeper, implying a high valuation of experience on the job. Alternatively, the expansion of higher technical education since the 1940s had produced an ample supply of new engineers which depressed the salaries of fresh engineers.[12] Further, if the 2002 data are accurate, one could note that relative earnings have almost returned to their level before 1940, and this despite a continuously high output of new engineers. One could speculate that young engineers were less in demand when production came on-stream within the old structure in the 1960s, and were more in demand when the new IT technology diffused in the 1980s.

With series on earnings and age profiles based on several benchmarks during the period of 135 years, the premium on higher technical education is estimated.[13] Besides the aforementioned data and the assumption about the graduation to civil engineer at the age of 25, the recipe followed here includes two more ingredients. One is the deflation of the earnings into constant prices of 1910/12 with the official consumer price index (CPI). The other is the discount rate, which is set constant at 2.7 per cent, reflecting the average historical real interest rate for government bonds during these 135 years. If instead the actual fluctuating rate were used, the result would be highly dependent on that factor and any meaningful interpretation distorted.

Figure 5.4 depicts the premium on higher technical education together with the schools' expenditures per student. The two curves strikingly mirror each other, the one being the reverse of the other. Any causality or interdependence between them should not be expected since the premium on education is neither an easily understood nor an easily available indicator, but derived from diverse factors which are more conceivable for actors to react to. However, *ex post* the premium on education could be one indicator of the success or failure of choices made by individuals and politicians.

To start with the perspective of the individual, a comparison with relative earnings in Figure 5.2 is illuminating. It must be remembered, however, that the earnings differential is a relative measure where an increase in the denominator, wages in manufacturing, is indicated by a decrease in the measure. The premium is, on the other hand, an absolute measure that discounts the value of the differential in constant prices. Keeping this in mind, one can note that the high relative earnings in the early part of the period were significantly modified over the life cycle. Relative earnings suffered a secular decline but the premium on higher technical education, or its profitability, was never

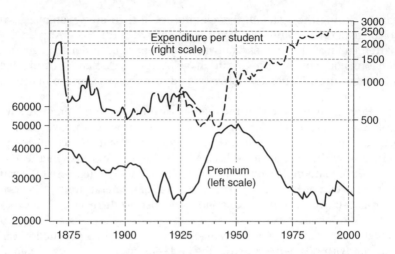

Figure 5.4 The premium on higher technical education and the expenditure of the schools per student 1867–2002, kronor in constant prices of 1910/12

Sources: See text and for details Ljungberg (2004b).

as high as during the 'big wave' in the 1940s and 1950s. The change from relatives to absolute figures plays a role, but age profiles of earnings are of course also important for the outcome. However, high earnings at young age do not only give a high gross income when discounted to the year of entry to working life, but also result in high forgone income during the years of study. Contributing to the 'big wave' of the premium was, first, low forgone income, and then the high relative earnings compared with college engineers until 1970. Over the very long term there is one dramatic change in the premium, the 'big wave' in the mid-twentieth century.

The late-twentieth century decline has a short break in the early 1990s. Actually that is the result of the introduction of new exams between the college engineer, who disappears, and the graduate engineer. Before the university diploma was introduced in 1989, and the bachelor of science in 1993, the study time for a graduate engineer had been prolonged from 4 to $4\frac{1}{2}$ years in 1986. Consequently the years of study that distinguish the graduate engineer from an engineer of lower rank first increased and then, a few years later, was reduced to $1\frac{1}{2}$ to $2\frac{1}{2}$ years. As a result the forgone income for a graduate engineer was roughly halved, which is seen in the jump of the premium in the early 1990s. It reflects a quality change in the involved parameters and a closer analysis would require more detailed data on the earnings for different categories of engineers. The relative decline in earnings of graduate engineers in the last decade might be a result of an upgrading of the lower engineers.

After the jump in the early 1990s, the premium resumes its decline. The latter change is of course dependent on the earnings structure of 2002, which is extrapolated into the future and to an increasing degree influences the premium after 1962. Nevertheless, it is an *ex ante* estimate on the basis of actual conditions. Moreover, a substantial change in the earnings differential is required to produce a significant difference. This is illustrated by the following counterfactual exercise: instead of a constant earnings differential over 2002–42 between graduate engineers and college engineers, let us assume an improvement within ten years back to the level of the year 1986 (compare Figure 5.2), and then remaining at that level until 2042. As a result the premium on education would have been 3.7 per cent higher in 1992, adding practically nothing to the long term change. In 2002 the premium would have been 5 per cent higher, thus only slightly improving the downward trend after 1994.

Output of higher technical education

From a social or political perspective, it might be interesting to note (in Figure 5.4) the jump in expenditure per student following the rise of the 'big wave'. The steady increase of expenditures after 1950 has, however, been mirrored by a steady decline in the premium on higher technical education. Is this a sign of diminishing returns, or are there other benefits that must be looked for elsewhere? A clue might be found in Figure 5.5, where the premium on higher technical education is added up to a total in each year, for all those graduating. Basically the same method was first used by Jorgenson and Fraumeni (1992), for a shorter period but with a much broader scope, as regards the education sector in the USA based on census income data for the total population, and including not only occupational income but also an estimate of non-market activity.

The premium on higher technical education has been discussed above, but the other component, the number of graduates, needs some clarification. Continuous data on graduates since 1956 are published in the official Statistical Yearbook (and in SCB 1996, for the period 1977/78–1994/95), but before 1956 the graduates are estimated from the numbers enrolled in higher technical education.[14] I have tried to take account of the aforementioned changes in the education of engineers through estimating the total of graduating in 'graduate engineer units' as a function of the total enrolled. Thus, from the closing years of the 1980s the output estimate loses in reliability, and since the earnings data of college engineers also changes with the appearance of the new university diploma and the bachelor of science degree, I have preferred to end the output estimate in 1992, which also is the last year of the expenditures account.

Two features are striking in Figure 5.5. The first, is that the output estimate is always above the conventional national accounts (NA) labour cost

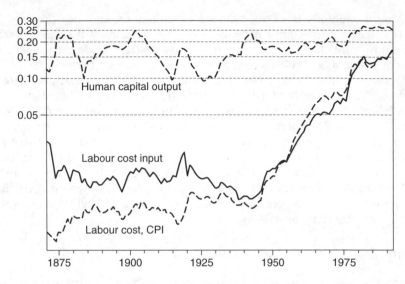

Figure 5.5 Estimates with different methods: contribution to GDP by higher technical education, 1871–1992 (percentage of GDP in constant prices of 1990/92)

Sources: See text and for details Ljungberg (2004b); GDP from SHNA.

estimate. Since the estimates are in constant prices of 1990/92, the level of the respective contributions is dependent on the relative prices in the early 1990s. With 1910/12 as a baseline, the output estimate would have changed hardly at all because the deflator for the premium on higher technical education (that is, the CPI) has over these 80 years changed by less than 1 per cent compared with the GDP deflator. On the other hand, the conventional NA labour cost estimate would have shifted further downwards, in 1910–12 crossing the CPI-deflated labour cost estimate. Two labour cost estimates are displayed for the purpose of illustrating the role of the deflator. The second feature that is striking is that the output estimate changes very little when measured as a share of GDP, in particular when compared with the labour cost estimates which increase very rapidly after 1940.

On the one hand, a possible conclusion to be drawn from Figure 5.5 would be that the national accounts under-estimate the contribution of higher technical education to GDP. On the other hand, a human capital estimate would only slightly increase the level of GDP, but it would decrease its rate of growth. The implication is that higher technical education has for many years had diminishing returns and declining productivity. Table 5.1 provides the rate of change for GDP and the different estimates of higher technical education (HTE). The conventional labour cost estimate, deflated by teachers' salaries, is actually an estimate of the input, although used in national accounts as a measure of the output (with zero productivity change

Table 5.1 National accounts and productivity of higher technical education (HTE): annual rate of change in per cent, 1871–1992

	1871–1940	**1940–92**	**1871–1992**
GDP	2.67	3.23	3.03
HTE, labour cost, conv.	2.73	9.03	5.26
HTE, labour cost, CPI	3.46	9.43	5.77
HTE, human capital	2.28	3.97	3.40
HTE Productivity I	−0.45	−5.06	−1.86
HTE Productivity II	0.73	0.40	0.51

Note: *Productivity I* = HTE human capital – HTE labour cost conv.
Productivity II = HTE labour cost CPI – HTE labour cost conv.

as a consequence). Until the mid-twentieth century the labour input in higher technical education just about kept pace with GDP growth, although in the second half of the twentieth century input increased almost three times as fast as economic growth. However, this comparison considers different aggregates, but if we find alternative output estimates for education, the difference between these and the input estimate will be a measure of its economic performance. Moreover, the difference in growth rates of the output and input estimates will be a measure of productivity growth. Thus, subtracting the conventional labour cost HTE estimate from the human capital estimate of the output results in 'Productivity I' : a dismal picture of diminishing returns.

However, it is not quite obvious that the micro level benefits of education could just be added up to a macro level estimate. First, if there exist institutional barriers in the labour market, for example a policy with the aim to compress the income distribution, then the earnings differentials would not reflect productivity differences and the lifetime income must under-estimate the human capital of professionals. However, I have argued elsewhere (Ljungberg 2004a) that the Swedish labour market at least since the First World War has been competitive enough for earnings largely to reflect the marginal productivity of labour. If institutional barriers thus are a less plausible explanation of the seemingly diminishing returns of higher technical education, then another explanation might be measurement problems. The macroeconomy is largely influenced by complementarities which could not be adequately expressed in value terms of the single parts.

The effect of using different deflators for the labour cost HTE estimate is illustrated in Figure 5.5. It is not self-evident that a labour cost estimate of non-marketed public services must be deflated with an index of salaries and wages. Seen from the view of the consumer, the tax payer, the expenditures have an opportunity cost in foregone private consumption, which is deflated by the CPI. Deflated in this way, a conventional labour cost estimate of public services could be said to reflect a valuation of the output of services made by the society through its political system. If this makes sense, 'Productivity II' can be computed as the difference between the CPI deflated 'output' and the salary index deflated input. A not so dismal picture emerges.

Conclusion

In the introduction the purpose of this chapter, to shed light on higher technical education in Sweden in relation to the problem of measurement of human capital, was stated. The analysis of the data makes it clear that since 1950, and at least until 1992, it is only through an increasing number of graduate engineers that the formation of human capital in higher technical education has kept pace with economic growth. From a theoretical point of view, diminishing returns on higher technical education have raised demands on ever increasing resources.

However, it might be that the conventional measures do not fully capture the content of human capital. Relative earnings of an engineer over a worker have declined drastically since the nineteenth century. The reason for this is not that the human capital in the engineer has decreased but actually that the worker has acquired a lot of human capital. 'Raw labour' is difficult to find nowadays. The theoretical problem is caused by the orthodox (Becker 1975) assumption that human capital is homogeneous (that is, that it can be substituted across different skills levels). However, more low-skilled labour can seldom substitute for high-skilled labour, and an equation built on this assumption is difficult to solve. Maybe the search for exact measures of some facts is futile, and we must be satisfied with different sets of data that throw light from different angles on the phenomena we study.

Finally, did higher technical education pay? The answer is not clear-cut. For the student, it probably did, and in particular for those graduating in the 1940s and 1950s. Thereafter the return has continuously been squeezed. For society the returns to higher technical education have been acquired with ever increasing costs. However, evaluating these costs, paid through taxes, at the opportunity cost of forgone private consumption indicates that higher technical education was beneficial for the economy. However, there might be spillover effects and complementarities that are not taken into account by the present estimates.

Notes

1   Financial support from the Bank of Sweden Tercentenary Foundation (RJ 95:156), and from the Swedish Council for Working Life and Social Research (2002:649) is gratefully acknowledged. I also thank Göran Ahlström, Anders Nilsson, Lars Pettersson and Lennart Schön for useful suggestions; however, the flaws are mine.
2   Records of Statens järnvägsbyggnader.
3   1867–1934, Riksdagsrevisorerna; 1924–1992, *Budgetredovisning*. In 1923 the fiscal year was changed from the calendar year to run July–June, and here calendar years are used throughout, the figures recalculated under the assumption that the expenditures have been equally divided between the two parts of the fiscal year.

4 A sevenfold increase is calculated from *Budgetredovisning*, yet apart from the government budget the public Bank of Sweden Tercentenary Foundation has (since 1968) been a main financier of research.

5 Estimated before 1929 from annual data on the expenditures and government's share thereof in 1929. Total expenditures were on average 18 per cent higher over 1867–1929, the size varying between 5 and 28 per cent depending on the distribution of expenditures between labour, investments and other costs.

6 Separate deflators have been constructed for labour cost, intermediate costs and investments, as reported in detail in Ljungberg (2004b).

7 Data for other education in Ljungberg (2004b).

8 Without going further into the issue of causes of economic growth here, it can be noted that Granger tests have shown that changes in labour productivity in manufacturing over 1890–1990 have persistently followed, with a time lag of up to a decade, upon changes in the enrolment in (all) higher education (Ljungberg 2002).

9 This housing allowance ceased in 1919 (Welin 1906, pp. 530 ff.; Statens järnvägar 1931, p. 118) when a system with a regional gradation of salaries and wages was introduced.

10 However, also within private business, graduate engineers earned very differently. Among the 19 branches organized in SAF, the average pay in 1953 exceeded that of the lowest paid branch with 33 per cent (Records of SAF).

11 Among the 86 graduate engineers at Kockums shipyard in 1961, the median age when taking the exam was 25, and the mean age 26.0. Quite a few (20 per cent) were under 24. In 1954 the graduation age of the 56 graduate engineers was marginally lower (average 24.8) (Records of Kockums). Obviously graduation age became higher: of 12,190 graduate engineers employed by SAF enterprises in 1977, some 149 were under 25; in 1986, the figure was 151 of 20,264 (Records of SAF). No efforts have, however, been made to correct for a change in the age of entry in the labour market; it is a welfare effect that probably also affected college engineers.

12 This alternative explanation seems less plausible. An internal memorandum of the managing board at Kockums shipyard, 25 June 1959, complaining about the scarcity of graduate engineers which could not be resolved by the substitution of college engineers, pointed out the particular requirements as 'general technical overview, ability to lead personnel, a good judgement and the capability for initiative and independent action'. Thus both technological knowledge and experience were in demand. Conferences with the other main shipbuilders and correspondence with SAF indicate that the memorandum reflected a common view (Records of Kockums). Pettersson (1983), in a penetrating study of college engineers, also characterizes the period 1954–66 as a period of excess demand for engineers. In a medium term perspective, relative salaries of young engineers increased in these years, as shown in Figure 5.3. However, in a long term perspective, relative salaries of young engineers were low, emphasizing the importance of experience and path dependency within the technological structure.

13 Benchmarks for age profiles are constructed for 1874, 1888, 1902, 1914, 1922, 1928, 1934, 1952, 1956, 1960, 1964, 1968, 1971, 1973, 1977, 1986 and 2002.

14 Estimated backwards, as an increasing function of those enrolled, to 1876 when graduates (in Stockholm) are given in Styffe (1877). Torstendahl (1975), using registers and catalogues (p. 291), shows a graph for graduates 1880–1910 (p. 85), and my estimate looks similar.

References

Ahlström, G. (1982), *Engineers and Economic Growth. Higher Technical Education and the Engineering Profession during the Nineteenth and Early Twentieth Centuries: France, Germany, Sweden and England* (London: Croom Helm).

Becker, G.S. (1975), *Human Capital*, 3rd edn (Chicago, IL: University of Chicago Press).

Budgetredovisning (for fiscal years 1920–93), (Stockholm: Riksräkenskapsverket/ Riksrevionsverket, 1921–93).

Jorgenson, D.W. and B.M. Fraumeni (1992), 'The Output of the Education Sector', in Z. Griliches (ed.), *Output Measurement in the Service Sectors*, Studies in Income and Wealth 56 (Chicago, IL: National Bureau of Economic Research), reprinted in D.W. Jorgenson, *Productivity. Volume 1: Post-war U.S. Economic Growth* (Cambridge, MA: MIT Press, 1995).

Levin, J. (1996), 'Government in the 1993 system of national accounts', in J.W. Kendrick (ed.), *The New System of National Accounts* (Boston, MA: Kluwer).

Ljungberg, J. (2002), 'About the role of education in Swedish economic growth, 1867–1995', *Historical Social Research*, 27 (4), pp. 125–39.

Ljungberg, J. (2004a), 'Earnings differentials and productivity in Sweden, 1870–1980', in S. Heikkinen and J.L. van Zanden (eds), *Exploring Economic Growth* (Amsterdam: International Institute of Social History).

Ljungberg, J. (2004b), 'Input and Output of the Swedish Education Sector, 1867–1995', unpublished manuscript.

Pettersson, L. (1983), *Ingenjörsutbildning och kapitalbildning 1933–1973* (Lund: Skrifter utgivna av ekonomisk-historiska föreningen).

Records of Kockums, GVII c: 3, 5, 13; KII a:28 (in Malmö stadsarkiv).

Records of SAF, F6 AI:107–15 (in Stockholms företagsminnen).

Records of Statens järnvägsbyggnader, D II:14–7 (in SJ Centralarkiv, Stockholm).

Riksdagsrevisorerna, 'Revisionsberättelse angående statsverket', *Bihang till riksdagens protokoll*, 2:a samlingen, 1:a avdelningen, 1:a bandet (Stockholm: annually 1867–1935).

SCB (1913–1993), *Statistisk årsbok.* Sveriges officiella statistik (Stockholm: Statistiska centralbyrån).

SCB (1996), 'Civilingenjörer i siffror', *Bakgrundsmaterial om universitet och högskolor 1996:2* (Statistiska centralbyrån).

SHNA: Swedish Historical National Accounts (Lund: Ekonomisk-historiska föreningen), in nine volumes:

    Krantz, O. (1986), *Offentlig verksamhet 1800–1980.*

    Krantz, O. (1987a), *Transporter och kommunikationer 1800–1980.*

    Krantz, O. (1987b), *Husligt arbete 1800–1980.*

    Krantz, O. (1991), *Privata tjänster och bostadsutnyttjande 1800–1980.*

    Ljungberg, J. (1988), *Deflatorer för industriproduktionen 1888–1955.*

    Pettersson, L. (1987), *Byggnads– och anläggningsverksamhet 1800–1980.*

    Schön, L. (1984), *Utrikeshandel 1800–1980.* Mimeo.

    Schön, L. (1988), *Industri och hantverk 1800–1980.*

    Schön, L. (1995), *Jordbruk med binäringar 1800–1980.*

SNA (1993), *System of National Accounts 1993* (New York: United Nations).

Statens järnvägar (1874–1951), *Lönestaten* (Stockholm).

Statens järnvägar (1931), *Minnesskrift. Statens järnvägar 1906–1931*, Vol. I (Stockholm: Kungl. järnvägsstyrelsen).

Styffe, K. (1877), *Kongl. Tekniska Högskolan i Stockholm* (Stockholm).

Torstendahl, R. (1975), *Dispersion of Engineers in a Transitional Society. Swedish Technicians 1860–1940* (Uppsala: Almqvist & Wiksell).

Tullkommitténs underdåniga betäkande af år 1882 (Stockholm).

Welin, G. (1906), 'Personalens löneförhållanden', in K.G.A. Welin (ed.), *Statens järnvägar 1856–1906*, Vol. IV (Stockholm: Järnvägsstyrelsen).

6
Economic Growth, Technology Transfer and Convergence in Spain, 1960–73[1]

Mar Cebrián and Santiago López

Introduction

Spain experienced an exceptional rate of economic growth between 1960 and 1973. This period, known as the 'Spanish miracle', was a major step towards closing the technology gap between Spain and other advanced countries.[2] The acceleration of growth in the majority of Western countries had begun a decade earlier, in 1950.[3] Spain, by contrast, did not develop significantly in the 1950s. Before 1960 the rate of growth of the Spanish economy fell short of the potential implied by the country's backwardness.[4] The post-1959 years saw a shift, resulting in a period of accelerated growth and significant 'catching-up'. Between 1959 and 1973 Spain had the highest rate of economic growth in Europe; among the member states of the OECD, only Japan enjoyed faster and more sustained growth (see Table 6.1).

In order to discover what the reasons for the Spanish economic growth were, we undertake an exercise of comparative growth accounting for this period in the next section.

Total factor productivity growth: an exercise in growth accounting

An exercise in growth accounting reveals the factors of production that contributed to growth in Spain, and the efficiency with which they were used. The results can also show the factors that contributed to the convergence between the economies of Spain and those of previously industrialized countries. The convergence discussion has shown that there are two processes required for income-convergence to take place: (a) attaining similar levels of capital intensity, and (b) attaining similar levels of technology. Although these two processes are inter-related, the first has received a good deal more attention.[5] Neo-classical growth theory assumes that technological progress is exogenous, accessible to all, and free, and thus it has devoted more attention to the issue of capital mobility. The rise of new growth theories has been, in part, a response to this abstraction, and these new theories have

Table 6.1 Levels and growth rates of GDP per capita, 1950–73

| Countries | Levels of GDP per capita, 1950–73 (1990 international Geary–Khamis $) | | | Annual percentage change of GDP per capita capita Growth rates | | GDP per relative to USA |
	1950	1960	1973	1950–60	1960–73	1960
Canada	7,437	8,947	13,838	1.85	3.35	0.79
France	5,270	7,543	13,123	3.59	4.26	0.67
Germany	3,881	7,685	16,966	6.83	3.41	0.68
Italy	3,502	5,789	10,643	5.03	4.68	0.51
Japan	1,926	3,988	11,439	7.28	8.11	0.35
Korea	770	1,105	2,841	3.61	7.26	0.10
Netherlands	5,996	8,289	13,082	3.24	3.51	0.73
UK	6,907	8,645	12,022	2.24	2.54	0.76
USA	9,561	11,328	16,689	1.70	2.98	1
Portugal	2,069	3,004	7,343	3.73	6.88	0.27
Spain	2,397	3,437	8,739	3.60	7.18	0.30

Source: Maddison (2001).

brought the issue of technology generation and diffusion to the forefront of mainstream economic research. Neo-classical economic theory has emphasized differences in factor endowments across countries and has devoted less attention to the possibility of differences in productivity and technology. However, empirical researchers have discerned that countries consistently differ in terms of productivity.

Recognition of technological differences by formal theory is stimulating empirical research on this issue, and in recent years there has been renewed interest in international comparisons of total factor productivity (Hulten 2000). A central theme of global economic change during the post-war period has been the application in less developed countries of technologies developed within economically-advanced regions. In addition, the post-war period has witnessed the convergence of per capita incomes of developed economies, and considerable catching-up by Japan and late industrializers such as South Korea. The 'technology gaps' among the industrial economies that typified the early post-war period have been narrowed by more rapid international technology transfer (Nelson 1990; Maddison 1995; Radosevic 1999).[6] One indicator of the contribution of innovative and technology-related activities to economic growth is the contribution to output growth of total factor productivity (TFP).[7] As Islam (1999) has pointed out: 'Of course, TFP differences are not identical to technology differences.[8] However, it is certain that technology difference leads to TFP difference and, in order to study the former, one must begin with the latter.'[9]

International differences in total factor productivity have been studied following the time-series growth accounting approach. Although the application of this methodology is not very common (due mainly to data constraints), we have employed this approach, applied by Jorgenson and his associates in the absolute form (Islam 1999, p. 493).[10] Thus we can provide TFP growth rates for each country, but not TFP levels. This methodology uses superlative indexes that exactly replicate a flexible representation of the underlying technology and their derived production functions, and considers the heterogeneity of capital and labour inputs. The methodology applied in this work uses the Divisia index (also called Translog or Törnqvist); this index fits a Translog production function.[11]

This methodology also allows us to analyse the sources of growth in real factor input between quantity and quality of factor inputs.[12] Simple measures of TFP do not adjust for the quality of labour and capital, and both are important. Particularly relevant is the contribution of the improvements in the design of new capital ('embodiment'), and the contribution of 'disembodied' technical progress, to economic growth. We consider that all inputs are different: one hour of work by an unskilled worker is not the same as one hour by a skilled worker. In conjunction with the literature on quality changes, we assign a role to embodied technical change as a determinant of the price of investment goods. This is done by estimating the service flow from different vintages of capital (Jorgenson 2001, pp. 11–12): that is, technological improvements in the design of investment goods (embodied technical change) may be a significant source of productivity change. The embodiment hypothesis implies that new capital is more productive than older capital (Hulten 1992, p. 965). This methodology considers that there are large differences in the marginal productivity of the different types of labour and capital. The Translog indices aid the disaggregate of the growth rates into quantity and quality growth rates. The importance of this distinction is that we assume that the introduction of new, more efficient capital goods and more qualified human capital are important sources of productivity change. If components with higher flows of capital input per unit of stock are growing more rapidly, quality will increase.

Two things are clear: first, that there has been much technological change in the production of new equipment; and second, that not all capital has the same quality. The production of capital goods becomes increasingly efficient with the passage of time. The failure to measure capital in efficiency units suppresses the quality effects into the conventional total factor productivity. Early accounting included in the residual not only pure (disembodied) innovation, but also the innovation embodied in capital goods (capital quality), human capital accumulation (labour quality), and improvements in markets (resource allocation). Recognizing the changes in the quality of capital is key to understanding the importance of technology transfer in the catching-up process by developing countries. For instance, if growth rates can be explained by improvements in the quality of capital, then the adoption of new machinery is an important factor in the success of Spain in this period. Conversely, if productivity improvement is relatively independent of factors of production, one must recognize the importance of disembodied technical change in Spain's economic development.

Methodology

The methodology is based on a constant returns to scale Translog production function, which provides a theoretical justification for the use of factor shares to weight growth rates (Christensen, Jorgenson and Lau 1973):

$$\ln Y = \alpha_0 + \alpha_k \ln K + \alpha_L \ln L + \alpha_t t + 1/2\beta_{KK}(\ln K)^2 + \beta_{KL}\ln K \ln L$$

$$+ \beta_{Kt} t \ln K + 1/2\beta_{LL}(\ln L)^2 + \beta_{Lt}\ln L \times t + 1/2\beta_{tt}t^2 \qquad (6.1)$$

where Y is output, K, L and t denote capital input, labour input and time, and where, under the assumption of constant returns to scale, the parameters α_i and β_{jm} satisfy the restriction:

$$\alpha_K + \alpha_L = 1$$
$$\beta_{KK} + \beta_{KL} = 0$$
$$\beta_{KL} + \beta_{LL} = 0$$

Denoting the price of capital by P_K, the price of labour input by P_L, and the price of output by P_Y we can define the shares of capital and labour inputs in the value of output, say v_K and v_L, by:

$$v_K = P_K K / P_Y Y$$

$$v_L = P_L L / P_Y Y$$

The necessary conditions for producer equilibrium are given by equalities between the value shares and the elasticities of output with respect to the corresponding inputs. Under constant returns to scale, the value shares for capital and labour sum to unity:

$$v_K = \frac{\partial \ln Y}{\partial \ln K}(K, L, t) = \alpha_K + \beta_{KK} \ln K + \beta_{KL} \ln L + \beta_{Kt} t$$

$$v_L = \frac{\partial \ln Y}{\partial \ln L}(K, L, t) = \alpha_L + \beta_{KL} \ln K + \beta_{LL} \ln L + \beta_{Lt} t$$

We can define the rate of productivity growth, say v_t, as the growth of output with respect to time, holding capital input and labour input constant:

$$v_t = \frac{\partial \ln Y}{\partial \ln t}(K, L, t) = \alpha_t + \beta_{Kt} \ln K + \beta_{Lt} \ln L + \beta_{tt} t$$

If we consider data at any two discrete points in time, say t and $t-1$, the average rate of technical change can be expressed as the difference between successive logarithms of output less a weighted average of the differences between successive logarithms of capital and labour input, with weights given by average value shares. The weights given to labour and capital are their relative shares in output. The rationale for this weighting system is the marginal productivity theory of distribution. This methodology assumes that factor prices are proportional to marginal product and that factor shares give a reasonable approximation of the elasticity of output with respect to each factor:

$$\ln Y(t) - \ln Y(t-1) = \bar{v}_K [\ln K(t) - \ln K(t-1)]$$

$$+ \bar{v}_L [\ln L(t) - \ln L(t-1)] + TFP_{t-1,t} \tag{6.2}$$

where

$$\bar{v}_K = \frac{1}{2}[v_K(t) + v_K(t-1)]$$

$$\bar{v}_L = \frac{1}{2}[v_L(t) + v_L(t-1)]$$

If aggregate capital and labour input are Translog functions of their components, we can express the difference between successive logarithms of aggregate capital and labour inputs in the form:

$$\ln K(t) - \ln K(t-1) = \sum \overline{v}_{Ki}[\ln K_i(t) - \ln K_i(t-1)]$$

$$\Rightarrow \text{Translog index of capital } \textit{input} \qquad (6.3)$$

$$\ln L(t) - \ln L(t-1) = \sum_j \overline{v}_{Lj}[\ln L_j(t) - \ln L_j(t-1)]$$

$$\Rightarrow \text{Translog index of labour } \textit{input} \qquad (6.4)$$

where

$$\overline{v}_{Ki} = \frac{[v_{Ki}(t) + v_{Ki}(t-1)]}{2} \qquad (i = 1, 2, \ldots, n)$$

$$\overline{v}_{Lj} = \frac{[v_{Lj}(t) + v_{Lj}(t-1)]}{2} \qquad (j = 1, 2, \ldots, p)$$

$\overline{v}_{ij}$ denotes the elasticity of each aggregate input with respect to each of its components subinputs or, again assuming perfect competition, the share of each subinput in total payments to its aggregate factor. These indexes adjust for improvements in the quality of aggregate capital and labour input by weighting the growth of each subinput by its average marginal product to a first-order approximation.[13] The quality index of aggregate capital is the difference between the Translog index of capital input (equation 6.3) and the Translog index of capital stock (that is, without weighting each subinput by its average marginal product).

Measuring factor supplies

Our analysis focuses on two aggregate inputs, capital and labour. We use data from the estimates of IVIE (Instituto Valenciano de Investigaciones Económicas) for the stock of capital. The depreciation rates are based upon the estimates of IVIE. We have compiled net capital stock estimates for different sectors: agriculture, fishing, energy products, metallic minerals and iron and steel products, agriculture and industrial machinery, electrical products, textiles products, food, beverage and tobacco, paper and printing, rubber products and others, wood, furniture and other manufactures, construction, insurance and credit institutions, transport and communications, and hostelry and restoration. In the case of labour, the task is to estimate the size and skills of the working population. Labour input is the product of employment, average hours worked, and hourly wage rates. The Wage Inquiry (Encuesta de Salarios) provides data about the average monthly wages of the number of employees per day and the hourly average monthly retribution for different groups: engineers, technicians with degrees, and graduates; technicians without degrees; administrative workers; officers, labourers and apprentices. With respect to the self-employed, data have been taken from *La renta nacional y su distribución provincial*, Banco Bilbao. One assumes that the self-employed earn a wage equal to the hourly wage of employees in the same industry. The agriculture salaries have been taken from Barciela, Carreras and Comín (1989).

Measuring factor shares

Spain has been recording data on aggregate compensation of employees as a percentage of GDP for several years.[14] To derive the aggregate share of labour, we multiply this estimate by the ratio of the total working population to paid employees. Under perfect competition and constant returns to scale, the aggregate share of capital is simply one minus this figure. To allocate capital input by asset type, we note that with geometric depreciation, and perfect foresight, the rental price of capital good K_i is given by:[15]

$$P_{Ki}(t) = P_{Ii}(t-1)r(t) + \delta P_{Ii}(t) - [P_{Ii}(t) - P_{Ii}(t-1)] \qquad (6.5)$$

where P_{Ii} denotes the investment price of capital good i, δ is the depreciation rate, and $r(t)$ is the nominal rate of return between periods $t-1$ and t. Operating under the assumption that all assets earn the same rate of return, we vary $r(t)$ until total payments to capital equal our estimate of the aggregate share of capital. This yields the estimate of the rental price of each asset category and, by extension, its share of payments to capital.

As a measure of output we have taken the GDP at factor cost given by Prados de la Escosura (1995).

Results

In this section a new estimation of total factor productivity growth and the rates of growth of real capital input and real labour input are presented. During the 1960–73 period, the highest growth rates of capital input were those of Japan, Germany, Korea and The Netherlands (see Table 6.2). Growth rates of capital input were the most significant source of output growth for the period 1960–73 in the cases of Canada, Japan, The Netherlands and the USA (see Table 6.3).[16] In all countries, the contribution of the growth of capital input exceeded the contribution of the growth of labour input. Furthermore, the contributions of growth in factor inputs were the predominant sources of economic growth for Canada, Japan, Korea, The Netherlands and the USA. According to this indicator of total productivity, the figures also reveal that growth in TFP was the most important source of growth in output by a substantial margin in Italy, Korea and Spain.[17] Spain's growth stemmed proportionally less from increases in labour supply and proportionally more from the growth of TFP than Korea's. However, the growth of capital input was more important in Italy than in Spain. In France, the UK and Germany, the growth of TFP was the most important source of growth; however, in these cases its contribution was very similar to that of the growth of capital input.

Table 6.2 Growth rates of GDP, factors and TFP, 1960–73

	Canada	France	Germany	Italy	Japan	Korea	The Netherlands	Portugal*	UK	USA	Spain†
Output	5.1	5.9	5.4	4.8	10.9	9.7	5.6	6.2	3.8	4.3	6.2
Capital	4.9	6.3	7.0	5.4	11.5	6.6	6.6	–	4.6	4.0	6.1
Labour	2.0	0.4	-0.7	-0.7	2.7	5.0	3.0	–	0.0	2.2	0.3
TFP	1.8	3.0	3.0	3.1	4.5	4.1	2.6	2.7	2.1	1.3	4.2

* 1965–73 for Portugal, see Amaral (2002)
† 1964–73. We have chosen the years 1964–73 for Spain because we have not found data about the capital stock and salaries for the period before 1964.
Source: Christensen, Cummings and Jorgenson (1980), except Spain.

Table 6.3 Contribution of growth rates of inputs and TFP to growth in real product (percentages), 1960–73

	Canada	France	Germany	Italy	Japan	Korea	The Netherlands	Portugal*	UK	USA	Spain†
Growth in real capital input	43.0	44.4	52.0	43.5	43.7	25.0	50.9	–	46.8	39.3	31.7
Growth in real labour input	20.9	4.3	-7.4	-9.0	14.7	32.9	3.1	–	-0.6	30.6	3.0
Growth in TFP	36.1	51.3	55.6	65.9	41.4	42.9	46.0	43.51	53.8	30.1	65.3

* 1965–73 for Portugal, see Amaral (2002)
† 1964–73
Source: Christensen, Cummings and Jorgenson (1980), except Spain.

Several previous growth accounting exercises conducted with diverse methodologies have focused on the importance of TFP to Spanish economic growth.[18] For example, the data offered by Suárez Bernaldo de Quirós (1992), who relaxed assumptions about constant returns to scale, exogenous technical progress and perfect competition, show that the contributions of labour, capital and TFP to growth between 1965 and 1974 were 6.5 per cent, 34.7 per cent and 58.8 per cent respectively. Dowrick and Nguyen's study (1989) argues that Spanish growth in the 1960s and early 1970s is explained by total factor productivity catching-up. This factor accounted for more than half of Spain's deviation from OECD growth.[19] We find a similar result in the studies carried out by De la Fuente (1995), Bajo and Sosvilla-Rivero (1995) and Raymond (1995). According to all of them, technological catch-up has been an important variable in Spanish GDP per capita growth.

The findings also illustrate the correlation between initial differences in levels of technology and relative rates of output growth. For example, countries such as Japan, South Korea and Spain, which were relatively backward, grew more rapidly (see Table 6.2). Moreover, the stronger TFP growth in these countries may reflect an effectiveness in technology transfer that compares favourably with the experiences of other developing countries.[20] The transfer of technology is not automatic, however, and not all countries have succeeded in closing the technology gap. One reason is that technology is not only public knowledge, but also has a component that is tacit (the provision of information not written down). This tacit element of technology makes the transfer more difficult.

The next step is to disaggregate the contribution of capital and labour input into the separate contributions of capital and labour quality, and of capital stock and hours worked. Quality changes in labour input are due to increases in the use of human capital, and quality changes in capital input are associated with embodied technical change (Hulten 1992). This method accounts for the differences in the marginal productivity of the different types of subinputs: that is, the rate of growth of each input between two years is a weighted average of the growth of its n components. Weights are determined by the share of each component in the total payments to its aggregate input. The basic objective is to divide capital and labour into homogeneous components, weighting the growth of each subinput by its average marginal product. The marginal products are measured by observed factor prices.[21] Labour quality reflects the substitution of workers with high marginal products for those with low marginal products. The improvement in capital quality represents the substitution towards assets with higher marginal products.

The results presented in Table 6.4 demonstrate that the growth of capital stock is very important to Spanish economic growth. The growth of human capital was instrumental to the increments of output in Italy, the USA, the UK and Spain between 1960 and 1973, but growth in the quality of capital

Table 6.4 Contribution of growth in quality of capital stock, quality of hours worked, capital stock, and hours worked to growth in GDP, 1960–73

	Canada	France	Germany	Italy	Japan	Korea	The Netherlands	Portugal*	UK	USA	Spain†
Growth of capital quality	9.7	0.5	3.7	3.1	11.4	10.2	15.3	–	4.1	0.96	1.9
Growth of labour quality	5.4	3.9	1.1	16.7	3.2	7.8	5.1	8.9	9.7	10.9	9.4
Growth of capital stock	33.4	36.0	49.0	39.9	32.4	14.8	35.2	45.3	42.7	28.9	29.8
Growth of hours worked	16.2	0.0	−11.1	−25.7	11.8	24.8	−2.0	−1.9	−9.7	19.1	−6.4
Growth of real factor input	64.7	40.5	42.7	34.1	58.8	57.6	53.6	56.5	46.8	59.8	34.7

* 1965–73 for Portugal, see Amaral (2002)

† 1964–73

Source: Christensen, Cummings and Jorgenson (1980), except Spain.

was of less significance in France, the USA, Spain, Italy, Germany and the UK. The two countries with the greatest increase in the quality of capital stock, as Table 6.5 shows, were Korea and Japan, where its contribution to economic growth was 10.2 per cent and 11.4 per cent, respectively (Table 6.4). The growth of labour quality was more important to the increment of output than the growth of capital quality in Italy, the UK, the USA, Spain and France. The contribution of the growth of hours worked was negative in Spain, as a result of the decline in the number of workers and their hours in obsolete branches of activity.[22]

We conclude that the principal sources of Spanish economic growth between 1964 and 1973 were TFP, capital stock and labour quality. In the Spanish case, however, there is a significant difference between the contribution of growth rates of inputs and TFP. The explanation regarding human capital has already been clarified, but with respect to the quality of the capital we cannot supply a satisfactory reason for the anomaly. The answer could be the Translog accounting method. This method assumes that higher prices of investment goods are equal to increases in the quality of capital. It means that factor prices are proportional to marginal productivity. A second supposition is that the improvements in capital quality are associated with embodied technical change, and represent the use of assets with higher marginal products. It is possible, then, to use price of capital as evidence of the marginal improvement in capital quality.

When an economy is technologically advanced, the Translog methodology takes account of the improvement in the growth of capital quality; the increase in the price of machinery can be considered a reliable estimate of capital quality. But in a backward economy, such as Spain in the late 1950s, this methodology transfers to the growth of TFP a part of the real improvement of the capital quality. For example, a firm buys a machine in an initial moment n. The productivity of this machine is 1 per hour of work and its price is 1. Now assume that, after a considerable period (20 years), a productivity gap has arisen between that firm and other firms on the technological frontier. Such was the experience of the Spanish economy during the 1940s and 1950s when it was isolated from global technological change. Further assume that the same firm buys a new machine after 20 years, on which productivity is 20 and its price is also 20. The difference between the Spanish firm and the firm of an advanced country is that the Spanish firm has to pay the price, 20, plus the cost of instruction/technical assistance in order to use the machine. The foreign firm only pays 20 for it, since it knows how to use the new machinery.

In the case of a significant technological gap even the Translog methodology only captures a part of the improvement of capital quality: that is, the price of that machine, which itself represents only one part of the total cost. The Spanish firm also pays for the 'know-how' in order to be able to use that new machine. Without this payment for technical assistance the

Table 6.5 Growth rates of capital stock, quality of capital stock, worked hours, and quality of worked hours (annual averages), 1960–73

	Canada	France	Germany	Italy	Japan	Korea	The Netherlands	Portugal*	UK	USA	Spain†
Quality of capital stock	1.1	1.2	0.5	0.4	3.0	2.7	2.0	–	0.4	0.1	0.4
Capital stock	3.8	5.1	6.6	5.0	8.5	3.9	4.6	2.7	4.2	3.0	5.7
Quality worked hours	0.5	0.4	0.1	1.3	0.6	1.2	0.5	0.9	0.6	0.8	0.8
Worked hours	1.5	0.0	–1.0	–2.0	2.2	3.8	–0.2	–0.1	–0.6	1.4	–0.5

Capital stock and worked hours are the arithmetic sum of subcomponents, with no adjustment for marginal productivity.
*1965–73 for Portugal, see Amaral (2002)
†1964–73

Source: Christensen, Cummings and Jorgenson (1980), except Spain.

Spanish company cannot engage in production, because it does not have the ability to operate the machine. This second payment is included in the TFP. This change of perspective requires a minor modification of the data. The result is that we are increasing the real contribution of TFP to Spanish economic growth. On the other hand the contribution of the quality of capital stock had to be higher than the figures show, because the payments made to acquire the know-how to use a higher quality capital are not included in the quality of capital. This correction would be more significant in the case of Spain, because its expenditure on the transfer of technology was, in relative terms, one of the highest in the world, as we will see in the next section.

Domestic innovation and technology import in Spain

The preceding section has demonstrated the importance of total productivity to Spanish economic growth in the period 1964–73. Increased efficiency was more important than the accumulation of factors of production. That efficiency required a certain degree of technical progress, which was achieved in two ways: by domestic innovation and/or by importing technology. The import of technology can be done by importing capital goods, by licensing, or by foreign investment. Given the scarcity of both R&D investment and high-level human capital in Spain, domestic innovation was made very difficult (see Tables 6.6 and 6.7; and López 1992). This is because technical knowledge is tacit and cumulative within individual firms; such tacit knowledge is neither machine-embodied nor codifiable, and thus is not easily transferable (Mowery and Rosenberg 1989; Pavitt 1993).

Thus we see in Table 6.6 that Spain has the lowest R&D expenditure in GNP.[23] If we look at levels of education in Spain, we see similar results.

Table 6.6 Total R&D expenditure as a percentage of GNP: Spain and other industrialized countries, 1962–71

Country	1962	1965	1967	1969	1971
USA	2.8	3.0	3.0	2.8	2.6
UK	–	2.3*	2.4	2.2	2.1[†]
France	1.4	2.1	2.2	1.9	1.8
West Germany	1.6	1.8	2.0	2	2.1
Japan	1.3	1.3	1.3	1.5	1.7
Italy	–	–	0.7	0.8	0.9
Portugal	–	–	0.2	–	0.4
Spain	–	0.14	0.2	0.2	0.3

* 1964
[†] 1970

Sources: Patrick and Rosovsky (1976), p. 533; for Italy OECD (1975); for Portugal Verspagen (1996); for Spain National Statistics Institute (*Instituto Nacional de Estadística*) (2000), and Prados de la Escosura (1995).

Table 6.7 Educational attainment of the working population 1964–73 (per cent)

Year	Illiterate	None and primary	Secondary or more
1964	7.21	85.38	7.42
1969	5.19	84.71	10.10
1973	5.28	79.09	15.62

Illiterate: those who are unable to read or write. None: those who have not finished primary school but can read and write. Primary: those who have attended elementary and primary schools. The length of this education may vary from 6 to 10 years. Secondary or more: those who have attended middle schools, secondary schools, high schools, universities and post-secondary professional schools.
Source: Mas, Pérez, Uriel and Serrano (1995), p. 72.

Table 6.7 shows that although the proportion of the population with a secondary education or better doubled between 1964 and 1973, it was still extremely low. This attests to a relatively poor level of human capital insufficient to innovate.

These low levels of human capital and R&D investment should have militated against the rapid economic growth and industrialization that took place in this period.[24] How did Spain manage to achieve such rapid progress, and take part in the process of catching-up, during those years? The decisive factor was its purchase of foreign technology. The import of technology was made possible when the economy opened and a new institutional context was established in 1959. In 1963, Spain paid $33 million for foreign licences and $11.3 million for technical assistance. In 1973, these figures jumped to $133 million and $105.1 million, respectively. Other economically backward countries, such as Japan and Korea, were able to narrow the gap by importing foreign technologies, although they also relied on two additional factors: high levels of human capital and significant expenditure in R&D. Spain, in contrast, has depended much more heavily upon on foreign technology to converge, and comparatively less on human capital and R&D investment. As a result, Spain has been able to catch up but has not become an innovative country like Japan or Korea. Unsurprisingly, Spanish expenditure on technology import though licensing and technical assistance contracts was higher than its R&D expenditure. Its payments for technology transfer amounted to 0.29 per cent of GDP in 1964, while investment in R&D was only 0.14 per cent of GDP. In 1973, these figures were 0.4 per cent and 0.31 per cent respectively. The emphasis on technology adoption compared with R&D investment suggests that the acquisition of foreign technology is responsible for the catch-up of the Spanish economy.[25] It certainly accounts for the expansion of Spanish production capacity during the period under study.[26]

Table 6.8 shows the relatively high level of Spain's technological dependency compared to other countries. In 1963, Spanish expenditure on

Table 6.8 Technological dependency, 1963–73: expenditures/incomes (royalties, copyrights and licences)

	1963	1964	1965	1966	1967	1968	1969	1970	1971	1972	1973
Spain (copyrights and patent royalties)	2.8	5.7	7.5	10.6	20.3	14.8	25.3	20.5	22.6	19.5	15.6
Japan (patent royalties)	16.5	10.6	10.9	9.9	8.9	10.2	7.6	7.5	7.7	7.7	8.1
Sweden (copyrights and patent royalties)	–	–	2.6	3.3	4.1	3.4	2.9	2.9	3.3	3.0	1.9
Italy (patent royalties)	–	–	3.5	3.7	3.1	3.4	4.2	4.0	3.9	5.6	5.0
Germany (licences and patents)	–	–	2.5	2.6	2.3	2.4	2.8	2.7	2.7	2.3	2.8
France (copyrights and patent royalties)	2.2	2.1	2.2	2.3	2.1	2.6	3.3	3.0	3.4	3.3	3.0
The Netherlands (royalties, licences, copyrights)	1.1	1.0	1.1	1.1	1.0	1.3	1.2	1.2	1.1	1.5	1.3
USA	–	–	–	–	–	–	0.2	0.2	0.2	0.2	0.2

Source: IMF, *Balance of Payments Yearbook*, vols 20–26, several years.

copyrights and patent royalties was 2.75 times higher than the amount it earned for selling its technology; by 1973 this figure had risen to 15.6.[27] In Japan, another country that spent more on technology acquisition than it generated, these figures were 16.5 and 8.1 respectively.[28] Growth can be attributed to technological innovation; because Spain produced few new inventions, this implies the transfer of technology from other countries. Access to innovation and technology allowed the income and productivity gaps to be reduced. In fact, this foreign technology import was common among many of the less developed countries, and is a major determinant of convergence between them and the developed countries.[29]

Technological catching-up requires country-wide capabilities (skills). For technology transfer to be effective, the acquisition of technological capabilities is necessary. How was Spain able to acquire technological capability and assimilate the technology imported given its low levels of education and scarce investment in R&D?[30] A possible answer is the high level of technical assistance supplied by foreign enterprises, and a transferred technology that was not very advanced, and thus relatively easy to adapt. Technology transfer was not merely an act of transferring proprietary information and rights from one firm to another. As Contractor (1985) has pointed out, technical services are often provided to facilitate the technology transfer. In fact, few licensing agreements provide patent rights alone; they include know-how and other forms of technical assistance. Given the level of Spanish human capital, intensive instruction from foreign suppliers was necessary for Spain to adopt imported technologies and to operate imported capital. We have calculated (thanks to technical assistance contracts) that the transfer costs needed to adapt the disembodied technology – that is, technical assistance payments – averaged 10 per cent of the total costs of the project (to construct or amplify a new industrial plant) and 23 per cent of the total payments in foreign currency. A great quantity of machinery was imported in order to apply the foreign technology (77 per cent of the total payments in foreign currency were to buy foreign machinery and 23 per cent to pay technical assistance). On the other hand, the simplest phases of the project were done in Spain by Spanish firms and paid in pesetas.

Figure 6.1 (technical assistance payments/royalties payments) illustrates the great significance of the quantity paid for technical assistance in relation to royalties. In an empirical study of technology transfers to 33 nations, Contractor (1980) showed that the payments for intellectual property rights divided by payments for administrative and technical assistance services increased as the industrialization and economic development of the recipient nation increased. Thus, as countries advance, they need proportionately less assistance in technical services and more extensive intellectual property rights. Spain, by contrast, increased its payments for technical services in the period 1963–71, as Figure 6.1 shows (mainly over 1963–66). Total payments for technical assistance and royalty payments were divided as follows: 25.4 per cent

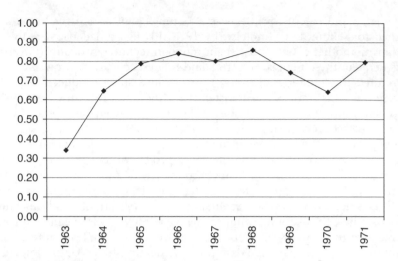

Figure 6.1   The ratio between costs of technical assistance and royalties, 1963–71

Source: Banco de España (1969).

of the total payments were for technical assistance and 75.6 per cent for royalty payments in 1961. By 1971, payments for technical assistance amounted to 44.3 per cent and royalty payments to 55.7 per cent. In other words, indigenous human capital was enhanced by foreign training.

The second reason behind the successful transplant of technology was the increased number of engineering and consultancy enterprises in Spain. The task of these enterprises was to apply the imported technology to the exigencies of the Spanish market. The Spanish government desired a return on their investment in foreign technology, and was keen that most of the engineering was carried out by these enterprises.[31] This type of enterprise did not account for the base engineering (this kind of engineering or know-how was imported), but only final engineering. These enterprises, which have been very important since the 1960s, were closely related to foreign enterprises. The great demand for techniques (due to industrial development) underlay the success of these enterprises. In 1972, there were 70 firms and 800 higher technicians (*Revista de Ingeniería Química*, August 1971 and July 1976).[32]

Third, the level of human capital required to understand the imported technology was not very high, since the technology transferred through licensing was not advanced. Licensing transactions transfer relatively mature technologies, easily codified products and process know-how. According to Mansfield *et al.* (1982), the average age of technologies transferred through licensing was 13 years more than those transferred through capital goods imports.[33] In the contracts we have studied for Spain, it is clear that the technology transferred through licenses was more mature than that which was transferred through capital goods imports. The case of technical assistance contracts is rather different. The technology transferred with this type of contract was much more advanced. The more complex technology of technical assistance agreements is reflected in a higher price of the technology transferred, and in a higher quality of the workers dispatched by the vendor company.

Conclusions

The results of this chapter suggest that total factor productivity growth in Spain was very high between 1964 and 1973, and that it was the decisive factor in the growth in output. This result is not surprising since, as convergence theory predicts, the flow of knowledge from technology leaders speeds the development of technology in the follower countries. In consequence, income per capita will grow faster as technology diffusion narrows the 'technology gap'. This process of catching up is not automatic, however. Closing the gap requires country-wide capabilities (skills), as well as firm-specific capabilities (know-how). Effective assimilation of foreign technology is complex. Less developed countries can increase their rate of growth when they are able to import and implement technologies invented abroad. Spain is a classic example.

The second point that should be underlined is how Spain was able to acquire technological capability and assimilate the technology imported given its low levels of human capital and limited investment in R&D. It overcame these handicaps by paying large amounts to foreign firms for technical assistance. The transfer of relatively unsophisticated technology permitted its assimilation with a low level of human capital. When the technology transferred was of a greater magnitude of sophistication, engineers had to be imported as well. Technical assistance played a key role in the expansion of Spain's productive capacity and technological capabilities. Much of Spain's competence thus came from the 'learning-by-doing' facilitated through licensing and technical agreements. Nearly all of the contracts signed by the Spanish firms provided assistance through the dispatch of personnel and the transfer of codified technical information, including blueprints of equipment designs. Spanish firms had few incentives to pursue technological accumulation through investment in R&D. Trade policy, and its persistent protectionism and limited competitive pressures, permitted the accumulation of production capacity with little or no accumulation of technological capabilities by domestic R&D.[34] In other words, advances in productivity in Spain were due to technology transfer rather than to investment in R&D.

Finally, we have shown that growth in the quality of capital stock was of minor significance in the Spanish economy. In the Translog methodology, price only captures a part of the quality of input capital in the countries not technologically advanced; that is, the price of the machine, which itself represents only one part of the total cost since it does not include the payments done for technical assistance. The quality of capital not only depends on the machinery but also on how you use it. When an economy has only a limited investment in R&D, as Spain did in the 1960s, it is necessary to solicit and pay for foreign technical assistance. This assistance is indispensable to operate the new machinery of higher quality. This payment is due to an improvement in the quality of capital and should be considered included in the quality of capital. However, this quantity is included in the TFP. What is actually a higher quality of the capital stock in our account therefore shows up in TFP. The bottom line thus showed too a big contribution of TFP, and too a small contribution of the quality of the capital stock to Spanish economic growth.

The findings in this chapter are consistent with a broad consensus obtained in other studies: that technology transfer can be an important source of productivity growth in developing countries when they are able to exploit the existing technologies available from leaders. This is particularly true when the technology transferred through contractual agreements was simple and highly standardized and, as a result, easy to understand. The potential for development in spite of the technology gaps that this model suggests is consistent with the Spanish experience between 1964 and 1973. During those years, Spanish firms secured substantial foreign technical assistance to assimilate foreign technology, and underwent tremendous economic development, almost 'catching up' with the developed countries by the end of the period.

Notes

1   We are grateful to J. Rosés, N. von Tunzelman, J. Ljungberg, B. van Ark, J. Reis, F. Comín, J. Simpson, L. Prados de la Escosura, G. Federico and participants at the Klinta workshop, as well as an anonymous referee.
2   The arithmetical average growth rate of advanced countries between 1950 and 1973 was 3.8 per cent per year, according to data from Angus Maddison (1983). Among developing countries, Spain, Korea, Singapore and Brazil had the highest growth rates: see Crafts (2000).
3   The autarchic period in Spain ended in 1959. Liberalization in Spain was gradual, but allowed remarkable acceleration after the stabilization plan in 1959.
4   As Prados de la Escosura and Sanz have said, 'the 1950s could be associated in Spain with incomplete catching-up, as the comparison with Germany and Italy tends to suggest': see Prados de la Escosura and Sanz (1995).
5   Bradford de Long and Summers (1991) have emphasized the importance of equipment investment. Mankiw (1995), for example, has argued that one can understand economic growth by focusing on education and capital accumulation, disregarding the determinants of technological progress. See Klenow and Rodriguez-Clare (1997) and Hall and Jones (1999) for a criticism of this position.
6   A large part of cross-country differences in income per capita are a result of TFP: see Easterly and Levine (2001).
7   This measure is not immune from criticism, owing to the restrictive assumptions involved. Furthermore, important sources of growth in TFP, such as economies of scale, cannot be derived by growth accounting techniques.

8 There was an early tendency to interpret the residual as technological progress. As Abramovitz (1993), has noted, the residual TFP measures the contribution of technological change to economic growth, but also scale economies or improvements in the efficiency with which resources are used. In the case of importers of technology TFP also includes technology transfer. Technological transfer implies not only the transfer of material goods, but also new ways of organization that can lead to improved efficiency and help to achieve economies of scale.

9 Islam (1999), p. 496.

10 The time-series growth accounting approach has been implemented in two forms, namely the absolute form and the relative form. For an explanation, see Islam (1999).

11 Growth accounting framework assumes the existence of an aggregate production function. A production function specifies a long-run relationship between inputs and outputs. There are a number of ways to approach the estimation of production functions and technical progress. The initial forms of production functions referred to the Cobb–Douglas, CES and the Leontief function. However, most of these functions have had restrictions imposed. On superlative indexes, see Diewert and Nakamura (1993).

12 The earliest growth accounts only took into consideration the physical quantities of the two main factors of production, capital and labour input.

13 For reasons of space the mathematical development of the Translog index is not complete; see Christensen, Cummings and Jorgenson (1980) and Jorgenson (1996).

14 Where necessary, we interpolate linearly between the state years to derive an estimate for the years of interest.

15 The rental price is higher if the depreciation rate is higher, due to more rapid physical deterioration.

16 The growth of private capital is also shown by Zabalza (1996).

17 It has not been possible to study the period 1960–73, as we do not have data for capital stock between 1958 and 1964.

18 The contribution of TFP to Spanish economic growth could be over- estimated, as the public sector is not completely included in the data. Research has shown that the productivity in the public sector in this period was inferior to that of the private sector (see Martínez Serrano and Muñoz 1995; Myro and Gandoy 1995; Pérez, Goerlich and Mas 1996; Sanchis Llopis 1997). The importance of TFP for the Spanish economy could also be explained by institutional change and spillovers during the period 1959–73.

19 These authors apply an extension of the classical model. Production is Cobb–Douglas augmented by a common rate of exogenous technological growth.

20 Dowrick (1992) has shown that the poorer OECD countries have experienced faster growth in the residual (total factor productivity) due to productivity catch-up. The ability of some developing countries to copy the techniques of production used by more advanced economies is the main reason for this productivity catch-up.

21 Zabalza (1996) has demonstrated that the growth rate of labour productivity in the period 1964–74 was 0.9 points lower than the growth rate of wages (labour costs are deflated by output prices). This result implies that our estimation about the growth rate of labour quality is over-stated, since we assume that wages are equal to marginal productivity.

22 See the drop of hours worked in Jaumandreu (1987), pp. 441–2.

23   Furthermore, basic research (the definition of basic research given in the Frascati Manual (OECD 1981), is: 'Work undertaken primarily for the advancement of scientific knowledge, without a specific practical application in view') accounted for a very low percentage of the national research effort. Applied research (R&D concerned with the application of technologies to the new products) and technological development (the translation of technical and scientific knowledge into concrete products and processes), were much more important. In 1967 Spanish firms devoted only 5.37 per cent to basic research, 36.44 per cent to applied research and 58.19 per cent of its expenditures in R&D to technological development. In 1973 these data were 3.02 per cent, 30.7 per cent and 66.21 per cent respectively (data taken from Instituto Nacional de Estadística 2000.) It is impossible to attain leadership in technology without maintaining a strong foundation in basic research.

24   There is an extensive literature demonstrating the contribution of R&D to productivity growth. See Abramovitz (1986), Fagerberg (1988; 1994), Coe and Helpman (1995), Verspagen (1994), Gittleman and Wolff (1995), Romer (1990) and Grossman and Helpman (1991). Education also has a significant impact on TFP: see Lucas (1988). Barro (1991) finds that low levels of educational enrolment are a substantial impediment to growth and to catching up.

25   The number of technology contracts rose from 246 in 1965 to 620 in 1972.

26   With regard to other types of technology transfer we have to say that the import of capital goods was also very important: $3,323.1 million between 1962 and 1966 and $5,421.5 million in the period 1967–71. Foreign direct investment was $417 million for the first period and $1,055 for the second one respectively.

27   These data do not include payments for technical assistance. If included, Spanish technological dependency would be much higher.

28   Japan's payments for imported technology substantial. As Ozawa has said: 'No doubt, technology imports have been the most significant stimulant to the development of Japan's own R&D industry.' For more on Japan's imports of technology, see Ozawa (1974) and Patrick and Rosovsky (1976). Japan, Mexico and India, among other countries, favoured licensing over investment: see Contractor (1985). The difference between Japan and Spain is that innovations developed in Japan were not wholly derived from imported technology, and were combined with intensive efforts in domestic R&D.

29   According to Abramovitz (1986), Baumol (1986), Grossman and Helpman (1991), Fagerberg (1994), Romer (1990), Parente and Prescott (1994) and Maddison (1995), technology transfer is a potential force for economic growth in backward countries.

30   We agree with those scholars who consider the process of catch-up to be very complicated, and one that involves a multitude of factors. Technology transfer includes not only the creation, transmission and reception of technical artefacts as end-products, but also the creation, transmission and reception of disembodied knowledge. As shown by Eicher (1999), von Tunzelmann (1995), Benhabib and Spiegel (2002), or Glass and Saggi (1998), the success of technology diffusion depends heavily upon the human capital levels of developing countries. Also see Barro and Sala-i-Martin (1995), who consider a model of costly imitation.

31   The Spanish government provided incentives to this type of firm after 1968. The Decree 617 of 4 April 1968 created the Register of engineering enterprises. Its objective was to make Spanish engineering enterprises more independent of foreign techniques, and ruled that currency payments for foreign engineering could not exceed 20 per cent of the total engineering import.

32 Also see Molero (1976), p. 61. The total number of people with a degree certificate working in Spanish firms in 1972 was 1,609 (see Mateo 1963).
33 Also see Contractor (1985) and Baranson and Roark (1985).
34 This conclusion has been taken from one of the chapters of Cebrián's doctoral thesis.

References

Abramovitz, M. (1986), 'Catching up, forging ahead, and falling behind', *Journal of Economic History*, 46, pp. 385–406.
Abramovitz, M. (1993), 'The Search for the Sources of Growth: Areas of Ignorance, Old and New', *Journal of Economic History*, 53 (2), pp. 217–43.
Amaral, L. (2002), 'How a Country Catches Up: Explaining Economic Growth in Portugal in the Post-War Period (1950s to 1973)', Doctoral thesis, European University Institute.
Bajo, O. and S. Sosvilla-Rivero (1994), 'Un análisis empírico de los determinantes macroeconómicos de la inversión extranjera directa en España, 1961–1989', *Moneda y Crédito, Segunda Época*, 194, pp. 107–36.
Banco de España (1969), 'Annual reports and balance of payments', *Boletín Estadístico*, December.
Baranson, J. and R. Roark (1985), 'Trends in North–South Transfer of High Technology', in N. Rosenberg and C. Frischtak (eds), *International Technology Transfer. Concepts, Measures and Comparisons* (New York: Praeger), pp. 24–44.
Barciela, C., A. Carreras and F. Comín (1989), *Estadísticas Históricas de España, siglos XIX y XX* (Madrid: Fundación Banco Exterior).
Barro, R.J. (1991), 'Economic Growth in a Cross Section of countries', *Quarterly Journal of Economics*, CVI (2), pp. 445–502.
Barro, R.J. and X. Sala-i-Martín (1995), *Economic Growth* (New York: McGraw-Hill).
Baumol, W.J. (1986), 'Productivity Growth, Convergence and Welfare: What the Long-Run Data Show', *American Economic Review*, 76, pp. 1,072–85.
Benhabib, J. and M. Spiegel (2002), 'Human capital and technology diffusion', *Proceedings*, December, Federal Reserve Bank of San Francisco.
Bradford de Long, J. and L.H. Summers (1991), 'Equipment Investment and Economic Growth', *Quarterly Journal of Economics*, CVI (2), pp. 445–502.
Christensen, L.R., D. Cummings and D.W. Jorgenson (1980), 'Economic Growth, 1947–1973, An International Comparison', in J.W Kendrick and B. Vaccara (eds), *NBER Studies in Income and Wealth*, Vol. 41 (New York: Columbia University Press), pp. 595–698.
Christensen, L.R., D.W. Jorgenson and L.J. Lau (1973), 'Transcendental Logarithmic Production Frontiers', *Review of Economics and Statistics*, 55, pp. 28–45.
Coe, D. and E. Helpman (1995), 'International R&D spillovers', *European Economic Review*, 39, pp. 859–87.
Contractor, F.J. (1980), 'The Composition of Licensing Fees and Arrangements as a Function of Economic Development of Technology Recipient Nations', *Journal of International Business Studies*, Winter, pp. 47–62.
Contractor, F.J. (1985), 'Licensing versus Foreign Direct Investment in U.S. Corporate Strategy: An Analysis of Aggregate U.S. Data', in N. Rosenberg and C. Frischtak (eds), *International Technology Transfer. Concepts, Measures and Comparisons* (New York: Praeger).
Crafts, N. (2000), 'Globalization and Growth in the Twentieth Century', *IMF Working Paper*, WP/00/44.

De la Fuente, A. (1995), 'Inversión, "catch-up" tecnológico y convergencia real', *Papeles de Economía Española*, 63, pp. 18–34.

Diewert, W.E. and A.O. Nakamura (1993), *Essays in Index Number Theory* (Amsterdam: North-Holland).

Dowrick, S. (1992), 'Technological catch up and diverging incomes: patterns of economic growth 1960–88', *The Economic Journal*, 102, pp. 600–10.

Dowrick, S. and D. Nguyen (1989), 'OECD Comparative Economic Growth 1950–85', *American Economic Review*, 79 (5), pp. 1,010–30.

Easterly, W. and R. Levine (2001), 'What have we learned from a decade of empirical research on growth? It's not factor accumulation: Stylized facts and growth models', *The World Bank Economic Review*, 15(2), pp. 177–219.

Eicher, T.S. (1999), 'Training adverse selection and appropriate technology: Development and growth in a small open economy', *Journal of Economic Dynamics and Control*, 23, pp. 727–46.

Fagerberg, J. (1988), 'Why growth rates differ?' in G. Dosi, C. Freeman, R. Nelson and S. Silverberg (eds), *Technical Change and Economic Theory* (London: Pinter).

Fagerberg, J. (1994), 'Technology and international differences in growth rates', *Journal of Economic Literature*, 32, pp. 1,147–75.

Gittleman, M.B. and E.N. Wolff (1995), 'R&D Activity and Cross-Country Growth Comparisons', *Cambridge Journal of Economics*, Vol. 19 (1), pp. 189–207.

Glass, A.J. and K. Saggi (1998), 'International technology transfer and the technology gap', *Journal of Development Economics*, 55, pp. 369–98.

Grossman, G. and E. Helpman (1991), *Innovation and Growth in the Global Economy* (Cambridge, MA: MIT Press).

Hall, R. and C. Jones (1999), 'Why do some countries produce so much more output per worker than others?', *Quarterly Journal of Economics*, 114, pp. 83–116.

Hulten, C. (1992), 'Growth Accounting when Technical Change is Embodied in Capital', *American Economic Review*, 82(4) (September), pp. 964–80.

Hulten, C. (2000) 'Total Factor Productivity: a short biography', Working Paper 7471 (Cambridge, MA: National Bureau of Economic Research).

Instituto Nacional de Estadística (2000), *La Estadística de I + D en España: 35 años de historia* (Madrid: INE).

International Monetary Fund, *Balance of Payments Yearbook*, Vols 20–26 (Washington).

Islam, N. (1999), 'International comparison of total factor productivity: a review', *Review of Income and Wealth*, 45 (4) (December), pp. 493–518.

Jaumandreu, J. (1987), 'Producción, empleo, cambio técnico y costes relativos en la industria española, 1964–85', *Investigaciones económicas (Segunda época)*, XI (3), pp. 427–61.

Jorgenson, D.W. (1996), 'Productivity and Economic Growth', in D.W. Jorgenson, *Productivity, International Comparisons of Economic Growth* (Cambridge, MA: MIT Press), Vol. 2, pp. 1–87.

Jorgenson, D.W. (2001), 'Information Technology and the U.S. Economy', *The American Economic Review*, 91 (1), pp. 1–32.

Klenow, P. and A. Rodriguez-Clare (1997), 'The Neo-classical Revival in Growth Economics: Has it gone too far?', *NBER Macroeconomics Annual*, 12, pp. 73–102.

López, S. (1992), 'Un sistema tecnológico que progresa sin innovar. Aproximación a las claves de la Tercera Revolución Tecnológica en España', *Ekonomiaz*, 22, pp. 30–55.

Lucas, R.E. (1988), 'On the Mechanics of Economic Development', *Journal of Monetary Economics*, 22, pp. 248–57.

Maddison, A. (1983), 'A Comparison of Levels of GDP per Capita in Developed and Developing Countries', *Journal of Economic History*, 43 (March), pp. 27–41.

Mar Cebrián and Santiago López   143

Maddison, A. (1995), *Dynamic Forces in Capitalist Development* (Oxford: Oxford University Press).
Maddison, A. (2001), *The World Economy. A Millennial Perspective*, (Paris: OECD).
Mankiw, N.G. (1995), 'The Growth of Nations', *Brookings Papers on Economic Activity*, 1, pp. 275–310.
Mansfield, E., A. Romeo, M. Schwartz, D. Teece, S. Wagner and P. Brach (1982), *Technology Transfer, Productivity and Economic Policy* (New York: W.W. Norton).
Martínez Serrano and J.A Muñoz, C. (1995), 'Sector Servicios', in J.L. García Delgado, (ed.), *Lecciones de Economía Española* (Madrid: Civitas), pp. 267–82.
Mas, M., F. Pérez, E. Uriel and L. Serrano (1995), *Capital Humano. Series históricas, 1964–1992* (Valencia: Fundación Bancaixa).
Mateo, J. (1963), 'El desarrollo tecnológico en España', in *Información Comercial Española*, May, pp. 35–44.
Molero, J. (1976), 'Las empresas de ingeniería', *Información Comercial Española*, August.
Mowery, D. and N. Rosenberg (eds) (1989), *Technology and the Pursuit of Economic Growth* (Cambridge: Cambridge University Press).
Myro, R. and Gandoy, R. (1995), 'Sector industrial', in J.L. García Delgado (ed.), *Lecciones de Economía Española* (Madrid: Civitas), pp. 245–66.
Nelson, R. (1990), 'US technological leadership: where did it come from, and where did it go?', *Research Policy*, 19, pp. 117–32.
OECD (1975), *Profils des Ressources Consacrées à la Recherche et au Developpement Experimental dans la Zone OCDE 1963–1971* (Paris).
OECD (1981), *The Measurement of Scientific and Technical Activities: Proposed Standard Practice for Surveys of Research and Experimental Development* (Paris: Frascati Manual).
Ozawa, T. (1974), *Japan's Technological Challenge to the West, 1950–1974: Motivation and Accomplishment* (Cambridge, MA: MIT Press).
Parente, S. and E. Prescott (1994), 'Barriers to Technology Adoption and Development', *Journal of Political Economy*, 102 (21), pp. 298–321.
Patrick, H. and H. Rosovsky (1976), *Asia's New Giant, How the Japanese Economy Works* (Washington, DC: The Brookings Institution).
Pavitt, K. (1993), 'What do Firms Learn From Basic Research?', in D. Foray and C. Freeman (eds), *Technology and the Wealth of Nations* (London: Pinter), pp. 29–40.
Pérez, F., F.J. Goerlich and M. Mas (1996), *Capitalización y crecimiento en España y sus regiones 1955–1995* (Madrid: Fundación BBV).
Prados de la Escosura, L. (1995), 'Spain's Gross Domestic Product, 1850-1993: quantitative conjectures', Working Paper 95-05, Universidad Carlos III, Madrid.
Prados de la Escosura, L. and J.C.Sanz (1995), 'Growth and Macroeconomic Performance in Spain, 1939–93', Discussion Paper 1104, Centre for Economic Policy Research, January.
Radosevic, S. (1999), *International Technology Transfer and Catch-Up in Economic Development* (Aldershot: Edward Elgar).
Raymond, J.L. (1995), 'Crecimiento económico, factor residual y convergencia en los países de la Europa comunitaria', *Papeles de Economía Española*, 63, pp. 93–111.
Revista de Ingeniería Química (1971; 1976).
Romer, P. (1990), 'Endogenous Technical Change', *Journal of Political Economy*, 98 (5), pp. 427–57.
Sanchis Llopis, T. (1997), 'Relaciones de intercambio sectoriales y desarrollo industrial. España, 1954–1972', *Revista de Historia Industrial*, 11, pp. 149–74.
Suárez Bernaldo de Quirós, F.J. (1992), 'Economías de escala, poder de mercado y externalidades', *Investigaciones Económicas*, 16 (3), pp. 411–41.

Verspagen, B. (1996), 'Technology indicators and Economic Growth in the European Area: Some Empirical Evidence', in B. van Ark and Nicholas Crafts, *Quantitative Aspects of Post-war European Economic Growth* (Cambridge: Cambridge University Press).

Verspagen, B. (1994), 'R&D and Productivity: A Broad Cross-section Cross-country Look' (Maastricht: MERIT Working Paper), pp. 93–007.

von Tunzelmann, G.N. (1995), *Technology and Industrial Progress: The Foundations of Economic Growth* (Aldershot: Edward Elgar).

Zabalza, A. (1996), 'La recesión de los noventa en la perspectiva de los últimos treinta años de crecimiento económico', *Moneda y Crédito*, 202, pp. 11–79.

7
Variations in Total Factor Productivity Analysed with Cointegration: Swedish Manufacturing Industry, 1950–94

Camilla Josephson

Introduction

This chapter aims to identify determinants for variations in total factor productivity (TFP). Not only have changes in productivity of Swedish manufacturing since the 1950s been sudden, but productivity on a disaggregate level has developed differently in various industries. Hence the causes of productivity variations may differ among industries with different methods of production. A comprehensive statistical analysis of three comparative data sets for labour-intensive, capital-intensive and knowledge-intensive industries in Sweden between 1950 and 1994 will explore this matter. The 'real-business-cycle theory' captures the leading explanation for short-term variations in TFP in which technological shocks constitute the key element (Black 1982; Kydland and Prescott 1982; Long and Plosser 1983; Prescott 1986). The most important macroeconomic evidence for such technological shocks is short-term variation in the Solow residual. In this study, TFP is calculated according to a Solow Cobb–Douglas production function in which the coefficients for labour and capital, both in logarithmic scale, are assumed to correspond to their marginal effect on TFP. This formula is used since, if certain conditions are met, it implies that TFP expresses shifts in technology (Solow 1957). These two conditions are perfect competition and constant return to scale. An alternative view is that TFP variations are caused by output movements arising from sources other than shifts in technology (that is, increasing or decreasing returns to scale would cause a Solow residual computed under the assumption of constant return to scale). This study aims to determine how violations of perfect competition and constant return to scale affect TFP variations in industries with different methods of production. Whereas the 'real-business-cycle theory' is consistent with the Walrasian equilibrium approach, the alternative view argues that equilibrium is unattainable and that variations in output are due to forces that cannot be explained by neo-classical economic theory. This study will explore the question of whether or not the determinants for TFP are consistent with the Walrasian equilibrium approach.

Statistical analysis within the growth context are frequently carried out by means of a regression model in which output constitutes the only dependent variable on the left-hand side, explained by a line of exogenous variables on the right-hand side. In contrast, this chapter uses the cointegrated VAR (vector auto-regressive) model. This is a dynamic model built on a full equation system able to distinguish between endogenous and exogenous variables in which all elements are allowed to change; hence there are no *ceteris paribus* or a priori assumptions involved. This model is able to do the following: (a) distinguish between short-term and long-term processes in the underlying structure; (b) describe the interactions and feedback effects within the system; and (c) describe the underlying driving forces. Thus cointegration analysis allows us to use a statistical model actively as a means of analysing the underlying processes that determine TFP.

An overview of the three different types of industries

Technology is here understood as a blueprint that combines production factors in different ways. The combination of labour and capital, on one hand, and the production technique, on the other, determines the methods of production in different industries. In other words, the quantitative use of each production factor, and how it is used in order to generate value added, defines the method of production. This study distinguishes among types of industries according to the production factor that receives the largest share of total factor remuneration. Swedish industries are classified as follows: wages to blue-collar workers are the largest factor of expenditure in *labour-intensive industry*; remuneration to capital constitutes the largest factor expense in *capital-intensive industry*; and salaries to white-collar workers constitute the largest factor cost in *knowledge-intensive industry*. Labour-intensive industries include earth and clay, wood, food and beverages, textiles and clothing, and metal manufacturing. Capital-intensive industries include steel and metal works, pulp and paper mills, and chemical industries. Knowledge-intensive industries include the engineering and printing industries. TFP has been calculated for the three subsectors with elasticities of capital and labour in each group derived from their respective share of value added. The elasticities of capital are 0.4 for labour-intensive, 0.6 for capital-intensive and 0.4 for knowledge-intensive industry.

The three sectors developed quite differently (see Table 7.1) with strong growth of production in the capital- and knowledge-intensive industries and weak growth in the labour-intensive industries. There are, however, notable differences between labour productivity growth and TFP growth. Labour productivity (Y/L) grew very strongly in the capital-intensive sector and rather strongly even in the labour-intensive sector. In both cases this

Table 7.1 Average annual percentage change of production, production factors, capital intensity, labour productivity and average TFP in industrial subsectors 1950–94

Sector	Y	K	L	K/L	Y/L	TFP
Labour-intensive	1.7	2.9	−2.3	5.1	4.0	1.6
Capital-intensive	4.5	4.1	−1.0	5.1	5.5	2.6
Knowledge-intensive	3.9	4.2	−0.3	4.5	4.2	2.5

Note: Y = output; K = capital; L = labour; K/L = capital intensity; Y/L = labour productivity; TFP = total factor productivity
Source: Computations on *SOS Industri*.

was accompanied by a rapid growth of the capital stock per working hour. Thus, their TFP growth was comparatively much reduced. In the knowledge-intensive industry capital stock grew more slowly which resulted in the smallest distinction between labour productivity and TFP among the sectors.

Description of the variables

Since this study aims to determine the effects of variations in capital intensity and of complementarities between labour and capital, we must look beyond labour productivity. Furthermore, if there are constant or increasing returns to capital and if labour carries out more skilled work, there must be increasing returns to capital and labour together and thus a rise in TFP (D. Romer 1996, p. 137). Research shows that capital investments, demand and workers' skills are believed to have a great impact on productivity variations, since they often constitute exogenous variables in models analysing economic growth (cf. Denison 1962; 1967; Jorgenson and Griliches 1967; Maddison 1989; Goldin and Katz 1998). Qualitative measures of labour and capital certainly affect TFP, but not necessarily in a unique manner. Accordingly this study analyses the relationships between TFP, machinery investments, export, electricity, and the share of white-collar workers with engineering skills, using a full equation system in which there are no a priori assumptions about which variables are exogenous and which are dependent. The variables are described below.

Total factor productivity

TFP growth is calculated in a growth-accounting procedure according to a traditional Solow Cobb–Douglas production function where the coefficients for labour and capital are in logarithmic scale:

$$Y = f(L^\alpha K^{1-\alpha})$$

$$\text{TFP} = d \ln Y - \alpha d \ln L - (1 - \alpha) d \ln K$$

Value added (Y) and capital stock (K) are in fixed prices and labour (L) in working hours, α (alpha) refers to factor shares for L and K, d implies first differences and ln stands for natural logarithm. All data are obtained from the national account statistics and the official industrial statistics. TFP has been calculated for three groups of industries in a similar manner with elasticities of capital and labour in each group derived from their respective share of value added. It should be noted that the share of labour is obtained by the actual price for labour in each industry in the same manner as in Translog growth accounting models, which implies that quality differences of capital and labour are not left in TFP. Labour elasticity of the subsectors is set at 0.6 for labour-intensive industry, 0.4 for capital-intensive industry and 0.6 for knowledge-intensive industry; consequently capital elasticity is set at 0.4 for labour-intensive industry, 0.6 for capital-intensive industry and 0.4 for knowledge intensive industry.[1] Varying the shares each year affects the result very little, so fixed shares have been used.

Machinery investments

The injection of new capital equipment is represented by machinery investment. This study emphasizes the importance of machinery equipment rather than total fixed capital as an indicator of new technology influencing productivity.

The share of white-collar workers with technical occupation

White-collar workers occupied in production and product development hold positions that imply a large degree of technical responsibility, so it is likely that these workers have higher technical education or training. Ideally, the measure of human capital should be the flow of new white-collar workers with engineering skills, but such data are not obtainable. The share of white-collar workers in the total number of workers occupied in production or product development here represents the accumulation of human capital. The share of white-collar workers in administration, marketing, management and so on is accordingly not accounted for. The share is only an approximation of human capital, particularly since the level of skills and the relative wages of blue-collar workers have risen considerably over the period. To avoid the direct impact of labour hoarding in business cycle downswings, the variable is transformed to an average of the white-collar share in the preceding three years. That is, the model estimates the impact on TFP of a build-up of white-collar competence so that only labour hoarding over longer periods of time is captured by the variable. Measuring education and experience among blue-collar workers is not uncomplicated since there is little available information about workers' skills. Although younger blue-collar workers have received more vocational training, older workers are more likely to be more experienced (Lundh 1991; Bengtsson 2003). The share of blue-collar workers increases during periods of expansion. Since it is shown that [an increase of employment] involves a rejuvenation of the labour force, it seems acceptable to interpret a decreasing share of white-collar workers as a proxy for falling age and sustained vocational training among blue-collar workers.

The export share

The export share of value added in the Swedish economy represents variations in the level of demand influencing economies of scale and capacity utilization. This measure can be used since Sweden is a small open economy with a high degree of export dependency in manufacturing. Export represents the most variable factor in demand. The export variable shows no significance in labour-intensive industry, since the output from labour-intensive industries is sold primarily on the domestic market. Export is thus not included in the data set for labour-intensive industry.

The electricity share of total energy

Electricity use in manufacturing is a variable with several connotations. In an earlier period, the electricity share would have been an indication of qualitative, technological change, since the substitution of electricity for fuel was part of the structural transformation of industry (Schön 1990; 2000). For the period after 1950 this is much less true. Electricity intensity is not an appropriate measure of capacity utilization since it is neutral to the export and we know that industries depending on economies of scale are large exporters and also large consumers of energy. Consequently, variations in export and electricity are only weakly correlated ($r = 0.22$). Due to fixed access to electricity, it cannot be increased on a short-term basis. The fact that decreasing electricity share often occurs simultaneously with rising TFP reflects the fact that more fuel is used in situations of high demand, particularly in energy-intensive industries. In this study, the share of electricity reflects short-term changes in demand.

Cointegration analysis: establishing economic steady state relations through analysing statistical equilibrium

Economic history is full of irregular and non-stationary paths. Since only stationary time series can be estimated in univariate Box–Jenkins techniques, the conventional wisdom has been to differentiate all non-stationary variables used in regression analysis. It is now recognized, however, that the appropriate treatment of non-stationary variables is not so straightforward in a multivariate context. Engle and Granger (1987) point out that a linear combination of two or more non-stationary series (series with a unit root) might be stationary. Where such a stationary 'I(0)' linear combination exists, the series are said to be cointegrated. The finding that more than 95 per cent of macro time series contain unit roots has spurred development of a theory of non-stationary time series analysis. A stationary linear combination of variables is called 'cointegration relation' or 'steady state relation', and constitutes a long-run equilibrium relationship. A vector error correction (VEC) model is a restricted VAR that has cointegration restrictions built into the specification, and is designed for use with non-stationary time series that are shown to be cointegrated. Since analysing cointegrated series by first difference entails a loss of information, the conventional wisdom on how to treat non-stationary time series has been revealed as inferior (Engle and Granger 1987; Enders 1995, p. 355; Johansen 1996; Johansen and Juselius 2000).

The basic objective of cointegration analysis is thus to find stationary linear combinations of variables containing unit roots. Cointegrated variables never move too far away from the steady state relation as they share a common stochastic trend. The cointegrated VAR model is a system of equations describing the time path of each variable. The most important feature of cointegrated variables is that their time paths are influenced by the extent of any deviation from the steady state relation. If the system is to return to the long-run steady state relation, the movements of at least some of the variables must respond to the effects of disequilibrium (that is, deviation from the steady state relation). The long-run steady state relation constitutes an attraction set towards which endogenous variables are drawn back after a shock in the system. For example, economic growth theory identifies a long-run relationship between output and capital investments. If a situation occurs in which the gap between output and capital investments widens considerably, relative to their long-run relationship, there are several ways to close the gap: (a) an increase in output or a decrease in capital investments; (b) an increase in capital investments with a commensurately greater increase in output; or (c) a fall in capital investments with a smaller decrease in output. Without a full dynamic specification of the particular model, it is impossible to determine which of the possibilities will occur. What we do know is that the short-term dynamics must be influenced by the deviations from the long-run relationship in order to re-establish the steady state relation. In an error correction model, the short-term dynamics of the variables in the system are influenced by the deviation from the steady state relation. Thus in order to determine the behaviour of each variable, we need to specify a model in which output and capital investment change in response to stochastic shocks and to the previous period's deviation from the steady state relation (see Enders 1995, pp. 365–6).

Equilibrium in a statistical sense refers to a stationary linear combination of variables in a model. Accordingly, if we include economically meaningful variables in a well-formulated statistical model, economic steady state relations can be identified from statistical equilibrium relations.

The cointegrated VAR model

In the present study, the main concern is to find steady state relations that determine variations in TFP; therefore an error correction model for this aim is specified. The empirical analysis follows the general-to-specific approach of Hendry and Mizon (1993) based on a VAR model with I(1) restrictions (cf. Johansen 1996). The variables (described in the previous section), are included in the variable vector X_t, defined as:

$$X_t = [Y_t M_t W_t E_t N_t]$$

where Y_t is total factor productivity (TFP), M_t is machinery investments, W_t is the share of white collar workers, E_t is export, and N_t is the electricity share of total energy, with t expressing time. All variables are in natural logarithms. The empirical analysis in this study applies to estimations based on the variable vector X_t using the cointegrated VAR approach. Some of the variables in the vector X_t contain a deterministic linear trend. A test confirms that a linear trend is significant. The vector error correction form used in this study is as follows:

$$\Delta X_t = \prod X_{t-1} + \sum_{i=1}^{k-1} \Gamma_i \Delta X_{t-i} + \mu_0 + \alpha \beta_0' t + \varepsilon_t$$

where $\prod$ can be decomposed as $\prod = \alpha \beta'$. α and β are two matrices of the dimension $(p \times r)$, where p is the number of variables and r is the cointegration rank (rank is the number of cointegration relations which is further discussed below). The β matrix represents cointegration vectors, describing how variables are cointegrated with each other within the steady state relation. The speed with which variables are drawn back to the steady state relation, or drift further away from the steady state relation after a shock in the system, is expressed by the α-coefficients. Γ_i are $p \times p$ dimensional matrices of autoregressive coefficients. $\mu_0 + \alpha \beta_0' t$, which are the only deterministic components in the model, imply that a constant is included and that coefficients are restricted to the trend term, so that a linear trend is permitted in both stationary and non-stationary directions. ε_t is the residual that, after diagnostic tests, is assumed to be independently and identically distributed. This precise model is used for estimations on branch-specific data for labour-intensive, capital-intensive and knowledge-intensive industries.

Determining lag length is the first step towards specifying a well-suited statistical model for each data set. The appropriate lag length is determined by analysing information criteria such as Hannan–Quinn and Schwartz. The number 1 is here chosen as the appropriate lag length in the VAR model for each industry. Naturally, before using the model it is important to test for any misspecification errors (see Table 7.2).

Since the interdependency of the residuals is critical to the analysis, the models are tested for autocorrelation by the Lagrange Multiplier (LM) test. The null hypothesis in the LM_1 and LM_4 test is that there is no serial correlation in the residuals of the first and fourth order, respectively. P-values above

Table 7.2 Misspecification tests of the model applied on each data set

Type of test	Labour-intensive industry χ^2	*p*-value
Autocorrelation LM_1	16	0.13
Autocorrelation LM_4	16	0.44
Normality	8	0.00
Type of test	**Capital-intensive industry** χ^2	*p*-value
Autocorrelation LM_1	25	0.12
Autocorrelation LM_4	25	0.28
Normality	10	0.28
Type of test	**Knowledge-intensive industry** χ^2	*p*-value
Autocorrelation LM_1	25	0.48
Autocorrelation LM_4	25	0.62
Normality	10	0.30

LM = Lagrange Multiplier.

0.05 imply that the hypothesis cannot be rejected and consequently that there is no autocorrelation in the residuals. Table 7.2 indicates that there is no autocorrelation in any of the residuals. A multivariate test of normality of the residuals is performed; for details regarding the test, see Hansen and Juselius (1995). The null hypothesis is that the residuals are normally distributed; *p*-values above 0.05 imply that the hypothesis cannot be rejected, whereas *p*-values below 0.05 indicate that the test is rejected and that the residuals are not normally distributed. The residuals in the model for labour-intensive industry are thus not normally distributed, whereas those for the other industries are. In terms of the analysis, the distributional form is of less importance than autocorrelation.

To determine rank implies identifying the number of cointegrated relations

The next step is to determine whether the time series are cointegrated, and if they are, to identify the number of cointegrating relations *r*. The software CATS implements VAR-based cointegration tests using the methodology developed by Johansen (1996). Johansen's method estimates the matrix in an unrestricted form, and then tests if we can reject the restrictions implied by the reduced rank of $\prod$. If we have *p* endogenous variables, each of which has one unit root, the linearly independent, cointegrating relations can range from zero to $p-1$.

The series may have non-zero means and deterministic trends, as well as stochastic trends. Similarly, the cointegrating equations may have intercepts and deterministic trends. Since the asymptotic distribution of the likelihood ratio (LR) test statistic for the reduced rank test depends on the assumptions made concerning deterministic trends, we must choose one of five possibilities considered by Johansen (see Johansen 1996, pp. 80–4, for details). The possibility that includes a constant and a linear trend in both stationary and non-stationary directions describes the data best, and is thus chosen. Now we can proceed sequentially from $r = 0$ to $r = p - 1$ until we fail to reject the hypothesis.

The trace test builds on eigenvalues λ_i. The magnitude of λ_i is an indication of how strongly the linear relation $\beta_i' R_{1t-1}$ is correlated with the stationary part of the process. When $\lambda_i = 0$ there is no error correction to the linear combination $\beta_i' R_{1t-1}$, which is non-stationary. If all λ_i, $i = 1, \ldots, p$, are zero, all linear combinations are non-stationary and there are no cointegration relations among the variables (Johansen 1996). The first column presents the eigenvalues, and the second column gives the p-values for the null hypothesis; the process can be made stationary in r numbers of directions. Using critical value at 5 per cent implies that p-values above 0.05 cannot reject the null hypothesis, and that the number of cointegration relations equals, or is larger than, the rank tested for. The lowest rank that fails to reject should be applied in the model. The trace test suggests rank 2 for labour-intensive industry. The number of cointegration relations in capital-intensive and knowledge-intensive industry is less certain, however. The p-values for one cointegration relation are 0.065 and 0.082, respectively. When the test results are uncertain, we can obtain more information by considering the number of roots inside the unit circle in the companion matrix, and the number of columns in which coefficients shows significant t-values in the α-matrix. α coefficients present speed of adjustment back to equilibrium; the first column in the α-matrix corresponds to the first cointegration relation, and thus we can expect significant coefficients in as many columns in the α-matrix as there are cointegration relations. Both capital-intensive and knowledge-intensive industries show two roots inside the unit-root circle, and only the first two columns have significant α coefficients; therefore rank 2 is chosen in both cases (see Table 7.3).

Table 7.3 Testing for cointegration of the sectoral series

Rank	Labour-intensive Eigenvalues	p-value	Capital-intensive Eigenvalues	p-value	Knowledge-intensive Eigenvalues	p-value
$r = 0$	0.80	0.000	0.80	0.000	0.79	0.000
$r \leq 1$	0.48	0.023	0.52	0.065	0.45	0.082
$r \leq 2$	0.28	**0.375**	0.30	**0.450**	0.37	**0.229**
$r \leq 3$	0.07	0.857	0.23	0.519	0.22	0.520
$r \leq 4$	*	*	0.10	0.718	0.11	0.672

* Labour-intensive industry has only 4 variables, so export is excluded.

Test for long run weak exogeneity

Linear combination of the variables Y, M, W, E and N that is stationary implies that the variables are cointegrated and, consequently, that their stochastic time paths must be linked. Not all variables in a stationary linear combination are affected by other variables, although they have an effect on these variables. Likewise, cointegrated variables can be affected by other variables, but not necessarily affect them. We can distinguish between *weakly exogenous* variables that have 'no levels feedback', and thus influence the long-run stochastic path of other variables without themselves being affected, and *endogenous* variables that are influenced by other variables, while affecting some of them. In order to determine whether a variable is exogenous, we analyse the α coefficient, which expresses variables' speed of adjustment to the long-run relations. If all adjustment parameters corresponding to a certain variable are equal to zero, the variable has a zero row in α (Juselius 2003, pp. 186–92). This implies that although the variable influences the long-run stochastic path of other variables, the original variable does not adjust to deviations from the long-run steady state relation, and should thus be considered as a common driving trend in the system. Hence exogenous variables that are included in the error correction model constitute a driving force for the whole system. Valuable information about the causal inter-relation among variables can be obtained by means of distinguishing between endogenous and exogenous variables. A likelihood ratio test for weak exogeneity is described in Johansen and Juselius (1990) and Johansen (1991).

The null-hypothesis is that the variables are weakly exogenous. The results are presented in p-values; a p-value above 0.05 means that the hypothesis cannot be rejected.

The test results for weak exogeneity, displayed in Table 7.4, reveal that there is no exogenous variable for labour-intensive industry. For capital-intensive industry export and the share of white-collar workers are strong

Table 7.4 Test for weak exogeneity

Type	TFP	Machinery investments	Export	White-collar workers	Electricity
Labour-intensive	Endogen. 0.00	Endogen. 0.00	Excluded	Endogen. 0.00	Endogen. 0.00
Capital-intensive	Endogen. 0.00	Endogen. 0.06	**Exogen.** 0.34	**Exogen.** 0.32	Endogen. 0.00
Knowledge-intensive	Endogen. 0.00	Endogen. 0.05	Endogen. 0.03	**Exogen.** 0.47	**Exogen.** 0.16

The null-hypothesis is that the variables are weakly exogenous. The results are presented in p-values; p-values above 0.05 mean that the hypothesis cannot be rejected.

candidates. In knowledge-intensive industry, the share of white-collar workers and electricity are strong candidates. However, we must determine if they are jointly weak exogenous variables. As the *p*-values are *0.37* and *0.35* respectively, the hypothesis of joint exogeneity cannot be rejected. In capital-intensive industry, exogeneity is imposed on export and the share of white-collar workers, whereas in knowledge-intensive industry, exogeneity is imposed on the share of white-collar workers and electricity.

Different causes for TFP in industries with diverse methods of production

The aim of the *real-business-cycle* approach is to find out whether a Walrasian equilibrium model provides a good description of the main characteristics of observed output fluctuations (Romer 1996 p. 151). Real business cycle analysis fundamentally investigates the role of neo-classical factors in causing economic fluctuations. Output fluctuations are viewed as arising from variations in the real opportunities of the private economy as opposed to monetary or nominal disturbances, therefore such approaches are known as *real-business-cycle* models or *RBC* models (Cf. King and Plosser, 1984). An important fraction of RBC economists use the Ramsey model to describe the aggregate economy, in order to construct a source for disturbances without shocks the model is extended to include changes in the production function from period to period. The change from one period to another is assumed to stand for shocks to the economy's technology (Kydland and Prescott 1982; Long and Plosser 1983; Black 1982). The principal macroeconomic evidence for the presence of technological shocks is the considerable short-term variation in the Solow residual (see Lucas 1975; Prescott 1986; Hansen and Wright 1992). There is an alternative theory, however, which argues that output fluctuations arising from other sources affect the measured TFP (that is, increasing or decreasing returns to scale would cause a Solow residual computed under the assumptions of constant returns to scale). Similarly, if industries use their production factors more intensively or efficiently when output is high, TFP computed under the assumptions of constant capacity utilization would rise as output increases (Romer 1996, p. 182; see also Bernanke and Parkinson 1991). This study aims to identify how violations of perfect competition and constant returns to scale affect TFP variations in industries with different methods of production. Such growth processes are only connected to technological change in the long run. In the short term they are connected to substitution and complementarity between labour and capital. Substitution means that one production factor substitutes for another (that is, capital substitutes for labour so that capital intensity per worker increases). Substitution sometimes causes decreasing returns to scale due to the necessity of changing input proportions when dealing with indivisible input goods. Complementarity between labour and capital occurs when an increase in the quantity or quality of one production factor increases the marginal return to the other factor, so that TFP increases. For example, improved quality in the labour force through training or higher education increases the return to both labour and capital so that total factor return increases and consequently TFP increases.

Earlier research shows that labour with specialist skills gained through work experience or through education constitutes a better complement to capital, while capital is a better substitute for unqualified labour (Welch 1979). It has been shown that blue-collar workers with vocational training and/or considerable work experience are more likely to increase the return to capital than blue-collar workers without these qualities (Lundh 1991; Lundh and Ohlsson 1994). These findings agree with the results of studies by Hirschhorn (1984) and Nuwer (1988). They have shown that since mass-production technologies integrate standardized operations into continuous-flow processes, workers need 'diagnostic skills' in order to ensure the continuity of production flows. The notion of 'diagnostic skills' is based on integrated production methods in which the decomposition of craft skills into component parts is argued to results in a separation of execution from conception. Such separation creates a wide gap between 'the experiential knowledge that comes from shop floor experience and the theoretical knowledge of machine fundamentals' (Hirschhorn 1984, p. 58). Diagnostic skills fill this gap.

However, it is also shown that white-collar workers with advanced engineering education constitute a better complement to new capital than white-collar workers with lower technical education and blue-collar workers (Pettersson 1983; Goldin and Katz 1998). The interpretation that white-collar workers form a better complement to new capital, whereas new capital will substitute for blue-collar workers to a greater extent, is thus not straightforward. The quality of labour is not the only important factor for complementarity. Just as important is the quality of capital, which can be measured by its age and physical contents. The capital stock consists essentially of machines and buildings. Since machines are used actively by labourers in production, whereas buildings merely accommodate machines and labour, it is likely that machinery investments have a more important influence on the complementarity between labour and capital.

Earlier statistical analysis within the economic growth context been carried out by means of growth accounting or regression models in which output constitutes the only dependent variable on the left-hand side, explained by a line of exogenous independent variables on the right-hand side (see Denison 1962; 1967; Kendrick 1973; Maddison 1989). Much has been learned through this method. Explanatory variables, such as quality indicators for labour and capital, influence output, but not necessarily in a unique manner. It seems more likely that machinery investments and the share of highly skilled workers are closely related. Moreover, it appears probable that the interdependency between such variables occurs differently in industries with diverse methods of production. New capital and labour may constitute a complementarity in one industry, whereas new capital might substitute for labour in another industry.

Identifying cointegration relations between TFP, machinery investment, export, the share of white-collar workers and the electricity share

Macroeconomic theory is useful for indicating possible steady state relations as well as proposing origins for external shocks. However, it cannot measure the coefficients within the steady state relation, or the size of the impacts from such shocks, and usually cannot determine whether the effects of external shocks are immediate or delayed. The statistical tools for improving our understanding of the underlying mechanisms and distinguishing between macroeconomic behaviour in the short, medium and long term, is provided by the cointegrated VAR model. By means of this model we can delineate inter-relationships between TFP, machinery investments, the share of white-collar workers, export, and the electricity share of total energy in the following ways:

(a) by testing whether a hypothetical steady state relation is empirically stationary (that is, if a linear combination of variables is shown to be mean reverting, we have identified a steady state relation);

(b) by estimating the β-coefficients describing the steady state relation, and thereby conceiving how variables are interrelated;

(c) by estimating the α-coefficients describing each variable's speed of adjustment towards the steady state relation, which gives us information on how dependent or independent individual variables are, and how they respond to stochastic shocks and to the previous period's deviation from the steady state relation;

(d) by gaining information on the origins of the shocks that push some variables away from the previous steady state relation (a similar definition is given in Johansen and Juselius 2000).

In order to learn more about the time series properties over the long term, and find out what type of economic growth process that might be reflected by the time series, this study tests eight different hypotheses for the inter-relations between TFP, machinery investments, export, the share of white-collar workers, and the electricity share, using the data set for each type of industry. Each hypothesis is formulated as a restricted linear combination of certain variables, which corresponds to a steady state relation supported by economic theory. In this test we impose restrictions on just one of the vectors, leaving the second vector unrestricted. The null hypothesis is that a linear combination is stationary and thus identifies a steady state relation; p-values above 0.05 imply that the hypothesis cannot be rejected. However, large p-values only confirm that the linear combination of variables is stationary. It is therefore important to interpret the coefficients' signs in order to determine whether the process reflects the economic behaviour presented in the hypothesis.

When the variables are in logarithms, as in this study, the coefficients constitute elasticities. To be able to interpret a cointegration relation as primarily connected to a particular variable, we need to normalize the former on the variable of interest (in this case, TFP). This procedure is similar to determining the dependent variable in a regression model. A key difference is that the choice of dependent variable in a regression model changes the estimates of the regression coefficients, whereas in a cointegration relation the ratios between coefficients are the same independent of the chosen normalization. In practice, normalization means setting a chosen variable X to unity and then dividing all variables' coefficients with the X-coefficient so that X becomes 1. Hence normalizing on TFP implies that all variables are divided by the TFP coefficient so that TFP becomes 1, which allows the interpretation of the proportional relation between the variables. A homogeneous relation between certain variables implies that their β-coefficients sum to zero (Juselius 2003 pp. 125–6). In order to determine the causal relation between TFP and the other variables in the homogeneous relation, we move TFP to one side of an equation, which means that we must alter the signs on all the other variables' coefficients. This manoeuvre should ensure that coefficients within a homogeneous relation sum to one. It should be noted that this is purely for demonstration, and that TFP is not the only dependent variable, but that all endogenous variables are dependent.

Since all industries have rank 2 we need to identify two cointegration relations: that is, two steady state relations in each industry. Testing the eight different hypotheses (H1–H8) carries this out.

(H1) Learning-by-investing: this means that diminishing returns are offset because knowledge creation is a by-product of investment. A firm that increases its physical capital learns simultaneously how to produce more efficiently. The hypothesis implies that the positive effect of workers' skills on productivity works through each firm's investment, and that an increase in a firm's capital stock leads to a parallel increase in its stock of knowledge (Arrow 1962; P.M. Romer 1986). Arrow derived the idea that knowledge and productivity gains come with investments and production from earlier empirical studies, for example by Wright (1936). Later, Schmookler (1966) showed that patents, as a proxy for learning, closely followed investments in physical capital. Restrictions for homogeneity between TFP and machinery investments that express a direct proportional relation between TFP growth and machinery investments test this hypothesis. In order to be interpreted as learning-by-investing, the share of blue-collar workers and machinery investments should increase.

(H2) Learning-by-doing: this implies that the labourer learns through experience, and that experience is obtained during the production process. It is considered an explanation for technological progress independent of the scale of production (Arrow 1962). Several empirical studies of the production process in various industries have shown a positive relation between current labour productivity and past cumulative output and investment (see Wright 1936; Hirsch 1956; Alchian 1963; Hollander 1965; Sheshinski 1967; Lieberman 1984). This study assumes that such a learning process has a particular effect on the relationship between machinery investments and blue-collar workers, and further that vocational training and work experience increases the level of learning-by-doing. Accordingly, learning-by-doing is assumed to have the largest affect on TFP in labour-intensive industry. The hypothesis is tested through restrictions for homogeneity between machinery investments and the share of white-collar workers. Since increases in TFP are due to experience obtained during the production process, decreases in the share of white-collar workers (which implies increases in the share of blue-collar workers), and increases in machinery investments should be relatively small compared to changes in TFP in order to interpret the process as learning-by-doing.

(H3) Economies of scale: if the average cost of producing any type of output under a given technology is increasing or decreasing with scale we say that there are economies or diseconomies of scale. The idea is related to increasing, decreasing and constant return to scale where an exact clone of a production process that lists all factors of production should double the output. Increasing, decreasing or constant return to scale is said to prevail if the output is greater than, smaller than, or equal to, twice the amount of output. This study applies Koopman's (1957) approach, which is that all cases of decreasing returns to scale are connected to some *indivisible commodity* in the surroundings. For instance, *indivisible input* means that the input in a specific capital good is indivisible in the sense that it becomes useless if physically divided. It has a maximal capacity, but can be under-utilized to produce less output. Another example concerns the *set-up cost* for preparing the tools needed and learning how to perform the tasks. Once the set-up cost is paid, the amount of output is proportional to the extra labour spent. Set-up cost is thus a form of indivisible 'readiness', as a 'half-ready' worker is useless. A third type, *specialization and Smithian division of labour*, implies that output per worker increases as the tasks are more efficiently divided, and that productivity increases as specialization becomes a deliberate strategy in accordance with an industry's specific abilities. When production decreases, however, the division of labour may not be the most efficient one (Koopmans 1957, p. 152; Arrow 1979). This study argues that capital-intensive techniques require large-scale production in order to cover major capital investments, and hence the return to scale increases as more capacity is utilized. Although machinery investments constitute one primary

condition to increase production scale, a necessary condition for large production series is a corresponding demand. The export share of total value-
added is therefore a good proxy for high capacity utilization. The hypothesis is tested through restriction for homogeneity between TFP, machinery
investments and export. Both machinery investments and export should
have positive signs in order to interpret the steady state relation as economies
of scale.

(H4) Lucas human capital growth model: models by Uzawa (1965) and Lucas
(1988) express the reproduction of capital as involving only human components, and no physical capital. The hypothesis is tested through restriction for homogeneity between TFP and the share of white-collar workers.
Accordingly, TFP growth is directly proportional to increases in the share of
white-collar workers.

(H5) Complementarity between subjective knowledge and capital: Lucas
(1988) and Azariadis and Drazen (1990) distinguish between 'subjective' and
'objective' knowledge: *subjective knowledge* is a rival good incorporated in
individuals as human capital, the use of which is exclusive; *objective knowledge* is based in equipment which can be used by anyone. This study recon-
structs the interpretation of Lucas's growth model in which complementarity
between subjective knowledge and physical capital constitutes the principal
mechanism generating TFP. The capital conception in the reformed model
involves both human components through subjective knowledge, and physical capital through objective knowledge. Given that subjective knowledge is
used actively within the production process, output in the reformed model
depends on three things: (a) the degree of new subjective knowledge that
is generated from the production process; (b) the disparity between subjective knowledge within a firm compared to competing firms, and the overall
knowledge within the economy; and (c) network effects, which are of great
importance in generating increasing factor return, which is shown by adding
either differences among consumers or stochastic choice processes (e.g., the
consumer must choose between two technologies such as Macintosh and
DOS, and to obtain equilibrium the fraction choosing a particular alternative must equal the fraction that has chosen that alternative in the previous
period). Accordingly, output in the reconstructed Lucas model depends on
the accumulation of new subjective knowledge, how long it remains a rival
good, and the advantages involved for consumers in choosing a certain good
depending on the number of users.

The share of white-collar workers and machinery investments here express
subjective and objective knowledge respectively. The hypothesis is tested
through restrictions for homogeneity between TFP, machinery investments
and the share of white-collar workers. The export variable is an indicator
for network effects; a large simultaneous increase in export reflects positive
network effects. Both machinery investments and the share of white-collar
workers should show positive signs in order to interpret the steady state
relation as complementarity between subjective knowledge and capital.

(H6) Negative sector shocks due to structural change and reallocation of labour: Lucas and Prescott (1974) propose a mechanism through which sectoral technology and relative demand shocks cause employment variations between industries. The idea is that reallocations of labour across sectors is time-consuming, and therefore employment falls more rapidly in sectors suffering negative shocks than it rises in sectors facing favourable shocks due to technological change. This study proposes a similar mechanism in which structural change reflects increased relative factor costs and decreased relative factor return, compared with both domestic industries using more advantageous production techniques, and with industries in low-income countries using the same method of production but having lower costs. Since capital investments will increase in sectors with increasing relative factor return, whereas capital investments decrease in sectors with falling relative factor return, this will give rise to fluctuations in labour demand between sectors. In highly industrialized countries, sectors with decreasing factor returns are often the most blue-collar intensive ones, which will be affected by labour hoarding (that is, cuts in the blue-collar share of the labour force). Hence, an increasing share of white-collar workers and decreasing capital investments seem to be good proxies for identifying negative sector shocks. The hypothesis is tested through restrictions for homogeneity between machinery investments, TFP and the share of white-collar workers; the share of white-collar workers should increase and machinery investments decrease. An increase in machinery investments would imply capital substitution for labour.

(H7) Reduced capacity utilization or diseconomies of scale: diseconomies of scale imply that duplicating all inputs yields less than twice the amount of output. Yet if all inputs are increased in the same proportion, there can be no barrier to replication, and so decreasing returns to scale can derive only from a fixed input or an input that cannot be increased in the same proportion as others. Thus the failure to double the output implies the presence of some indivisible input, not listed among the arguments in the production function, which prevents replication (Sraffa 1925; 1926; Young 1928; Kaldor 1966). Negative effects of reduced capacity utilization on TFP are consequently connected to substitution for indivisible goods. Decreased capacity utilization involves decreasing return to capital and higher costs per unit of output. The hypothesis is tested through restrictions for homogeneity between machinery investments, export and the share of white-collar workers. In order to reflect reduced capacity utilization which gives rise to diseconomies of scale, there should be a decrease in export and machinery investments, while the share of white-collar workers should increase.

(H8) Reversed complementarity due to path dependency and lock-in effects: path dependency implies that we build on what we have, that a one-chance experimentation with technology leads to additional experimentation with

that same technology, which increases its advantages over untried alternatives. Accordingly, complementarity between subjective knowledge and
capital implies increased factor return in an initial stage. However, as economic prerequisites such as consumer choice processes, diffusion of new
technology, and external network effects change, path dependency leads to
lock-in effects. The more rigorously that knowledge grows in a single direction, the better are the chances for competitive advantages and increasing
factor return, although the risks for lock-in effects in the long run are also
higher. Decreasing factor return arises when knowledge continues to grow
in the same direction, despite the fact that the use of the same production
factors in alternative ways, or the use of other production factors for the
same purpose, would lead to higher factor return. (See Table 7.5.)

The ability to optimize investments depends on the quality of the information about the future. Since a second set of circumstances is always possible,
it is likely that an *ex-ante* efficient decision may not turn out to be efficient
ex post (that is, in retrospect), which is shown by Coase (1964), Calabresi
(1968), Demsetz (1969), Dahlman (1979) and Williamson (1993). The literature on path dependency, both theoretical and empirical, contains a number
of claims that path-dependency processes lead to inefficiencies, even for
products sold in open markets; the most important are David (1985), Arthur
(1989; 1990), Katz and Shapiro (1986), and Liebowitz and Margolis (1990;
1994; 1995; 1996).

Once the level of subjective knowledge within a firm equals the average
level of knowledge within the industry on an international level, the marginal cost of using highly skilled labour in high-wage countries will exceed
its marginal product value. On the other hand, low-wage countries have the
chance to compete with more developed countries due to the diffusion and
standardization of technology. When competition increases due to diffusion of knowledge and technology, the positive effects of complementarities

Table 7.5 The hypotheses in model form

Hypothesis	Test formulation
H1 Learning by investing	$\beta_1(Y-M)+\beta_2 E+\beta_3 W+\beta_4 N+\beta_5 T \sim I(0)$
H2 Learning by doing	$\beta_1 Y+\beta_2(M-W)+\beta_3 E+\beta_4 N+\beta_5 T \sim I(0)$
H3 Economies of scale	$\beta_1(Y-M)+\beta_2(Y-E)+\beta_3 W+\beta_4 N+\beta_5 T \sim I(0)$
H4 Lucas growth model	$\beta_1(Y-W)+\beta_2 M+\beta_3 E+\beta_4 N+\beta_5 T \sim I(0)$
H5 Complementarity between subjective knowledge and capital	$\beta_1(Y-M)+\beta_2(Y-W)+\beta_3 E+\beta_4 N+\beta_5 T \sim I(0)$
H6 Negative sector shocks	
H7 Reduced capacity utilization	$\beta_1 Y+\beta_2(M-E)+\beta_3(M-W)+\beta_4 N+\beta_5 T \sim I(0)$
H8 Reversed complementarity	$\beta_1 Y+\beta_2(M-E)+\beta_3 W+\beta_4 N+\beta_5 T \sim I(0)$

Y = TFP, M = machinery investments, E = export, W = the share of white-collar workers, N = the
electricity share of total energy, and T is the trend, I(0) implies a stationary state.

between highly skilled labour and capital vanish and this creates lock-in effects. Reversed complementarity between labour and capital implies that an increase in the quantity or quality of one factor decreases the jointly marginal return to both production factors, which results in decreasing TFP. The hypothesis is tested through restrictions for homogeneity between machinery investments and export. Machinery investments should increase although export is decreasing. In addition, there should be an increase in the share of white-collar workers in order to reflect reversed complementarity due to path dependency and lock-in effects.

Testing for H5 and H6 has identical restrictions. However, we can easily distinguish which of the hypotheses is confirmed when the coefficients are estimated in Table 7.6, since machinery investment should have positive sign in H5 and negative sign in H6.

Interpreting the results

Labour-intensive industry

The restrictions for H1 expressing learning-by-investing cannot be rejected. Since there is a proportional 1 to 1 relation between machinery investments and TFP, and a relatively large increase in the share of blue-collar workers, this is a case of increased throughput and/or increased scale. This does not necessarily mean that the process reflects learning-by-investing, but may reflect the fact that industries use their production factors more efficiently when output increases, so that TFP computed under the assumptions of constant capacity utilization rise as output increases. This study combines Chandler's (1962) arguments on increased throughput versus increased capacity, and the findings on diagnostic skills made by Hirschhorn (1984) and Nuwer (1988), in order to interpret the results for H1. In order to increase throughput without increasing scale, the integration of standardized operations into continuous-flow processes must be extremely smooth. Workers' 'diagnostic skills' are thus very important in order to ensure the continuity of production flows. This study argues that the effect of workers' diagnostic skills on factor return increases as output increases. In this H2 cannot be rejected. In order to reflect learning-by-doing, the increases in machinery investments and blue-collar workers should be quite small in relation to the change in TFP, which indeed is the case. Where TFP increases by 1, machinery investments and the share of blue-collar worker increase by 0.13 each. As the export variable is excluded, H3, H7a and H8 cannot be tested. The restrictions for H4 expressing the Lucas human capital growth model cannot be rejected, but since the coefficient for machinery investment is negative it does not reflect capital accumulation but rather labour hoarding. The restrictions for H5 cannot be rejected but the signs of the coefficients do not confirm the hypothesis of complementarity, and neither do they confirm capital substitution for labour as machinery investment decreases.

Table 7.6 Results of testing the eight different hypotheses expressing steady state relations on data sets for labour-intensive, capital-intensive and knowledge-intensive industry

Labour-Intensive Industry

Hypothesis	TFP	Machinery investments	Export	W	N	t	p-value
H1	1 =	+1		−2.27	+0.18	0	0.27
H2	1 =	+0.13		−0.13	−0.04	0	0.88
H3*							
H4	1 =	−0.41		+1	−0.10	0	0.07
H5, H6	1 =	−1.30		+2.30	0	0	0.26
H7*							
H8*							

Capital-Intensive Industry

Hypothesis	TFP	Machinery investments	Export	$W^\dagger$	N	t	p-value
H1	1 =	+1	+0.50	+4.23	−2.67	0	0.73
H2	1 =	−0.11	−0.21	+0.11	0	+0.01	0.86
H3	1 =	+0.68	+0.32	+3.11	−1.89	0	0.62
H4	1 =	+0.14	+0.05	+1	−0.51	0	0.07
H5, H6:	1 =	+0.16	+0.11	+0.84	−0.55	0	0.04
H7	1 =	−0.10	−0.15	+0.25	0	+0.01	0.28
H8	1 =	+0.03	−0.03	+1.01	−0.39	0	0.02

Knowledge-Intensive Industry

Hypothesis	TFP	Machinery investments	Export	$W^\dagger$	$N^\dagger$	t	p-value
H1	1 =	+1	+0.37	−2.43	−0.66	0	0.00
H2	1 =	−1.17	−1.14	+1.17	0	+0.07	0.06
H3:	1 =	+0.45	+0.54	+0.53	0	−0.04	0.75
H4	1 =	+1.15	+1.38	+1	0	−0.09	0.58
H5, H6	1 =	+0.46	+0.53	+0.54	0	−0.04	0.82
H7	1 =	−0.57	−0.47	+1.04	0	+0.03	0.02
H8	1 =	+0.03	−0.03	+0.38	−0.13	0	0.73

Export is excluded in labour-intensive industry
* The hypothesis cannot be tested since the export variable is missing
$\dagger$ The variable is exogenous, 0 implies that the coefficient is non-significant and thus set to zero.
W = the share of white-collar workers;
N = the electricity share of total energy and t is the trend.

It is, on the contrary, a clear indication of labour hoarding. H6 tests for negative sector shocks expressed as labour hoarding and decreasing machinery investments, which cannot be rejected, and the coefficient signs are in accordance with the hypothesis. Thus, learning-by-doing and negative sector shocks constitute the two steady state relations in labour intensive industries.

Capital-intensive industry

Although H1 cannot be rejected, the increase of white-collar workers and export does not reflect learning-by-investing, but is a clear indication of economies of scale and capital substitution for labour. Restrictions for H2 cannot be rejected, yet the signs of the coefficients do not confirm learning-by-doing but labour hoarding and decreasing return to scale. H3 cannot be rejected, and since the sum of the increases in machinery investments and export leads to a directly proportional increase in TFP, the interpretation is straightforward; the process reflects economies of scale. The simultaneous increase in the share of white-collar workers expresses capital substitution for labour, and the rather steep decline in the electricity share reflects increased use of fuel when production rapidly increases. H5, expressing complementarity between subjective knowledge and capital, and H6, expressing negative sector shocks, are both rejected. H7 cannot be rejected, and since machinery investments and export decrease simultaneously as there is a relatively large increase in the share of white-collar workers, the hypothesis reflects labour hoarding and diseconomies of scale. H8 (testing for reversed complementarity) is rejected.

Economies and diseconomies of scale thus constitute the two steady state relations in capital-intensive industry. In testing the eight hypotheses, however, it emerged that not only H3 and H7 expressed economies and diseconomies of scale respectively, but that H1 and H2 did as well. Therefore capital-intensive industry has two models. Model 1 includes the steady state relation expressed in H1 and H2, whereas Model 2 includes the steady state relations expressed in H3 and H7. Export and the share of white-collar workers are exogenous variables in capital-intensive industry. Exogenous variables influence the long-run stochastic path of the other variables, without being affected themselves. Accordingly export, which here expresses demand, constitutes a common force driving the entire system in capital-intensive industries. An increasing share of white-collar workers may reflect how heavy machinery investments substitute for blue-collar workers. However, knowing that decreasing returns derive from the unfeasibility of changing input proportions of indivisible production factors, changing the share of white-collar workers can also express diseconomies of scale.

Knowledge-intensive industry

H1 and H2 (expressing learning-by-investing and learning-by-doing) are clearly rejected. H3 expressing economies of scale cannot be rejected, which implies that changes in TFP are proportionate to the changes in machinery investment and export. H4 tests for Lucas endogenous growth model, which cannot be rejected. Since machinery investments increase relatively more than the share of white-collar workers, the process does not only reflect the effect that human components have on TFP, as stated in the hypothesis. H5 expresses complementarity between subjective knowledge and capital, which cannot be rejected. The estimates of the coefficients for machinery investments, export and the share of white-collar workers are very similar in H3 and H5. However, knowing that the latter is more widely accepted than the former, comparing the coefficients in the two models might generate further conclusions.

We see that machinery investment has a slightly larger effect on TFP in H3 compared to H5, since a smaller change in machinery involves the same changes in TFP. A slightly larger increase in export in H3 compared to H5 produces the same effect on TFP, whereas a slightly smaller increase in the share of white-collar workers in H3 compared to H5 produces the same effect on TFP. The conclusion is that increases in machinery investments and the share of white-collar workers affect TFP slightly more than increases in export. However, the fact that the export variable is large and significant reflects the importance of network effects. H6 (expressing negative sector shocks) and H7 (declaring reduced capacity utilization) are both rejected. H8 cannot be rejected and, since the signs are in accordance with the hypothesis (that is, machinery investments and the share of white-collar workers increases as exports decrease), the process reflects reversed complementarity due to path dependency. Accordingly, complementarity processes between subjective knowledge and capital, and reversed complementarity due to path dependency and lock-in effects, constitute the two steady state relations in knowledge-intensive industry. The share of white-collar workers and electricity are exogenous variables. Since electricity is not significant in all steady state relations we cannot regard it as a common driving force in the system. The only identifiable common driving force is the share of white-collar workers, which fits with H5 and H8 since it express the level of subjective knowledge.

Interpreting α (alpha) and β (beta) coefficients

Both cointegration relations have to be jointly identified before the alpha and beta coefficients can be interpreted. Variables' roles in the cointegration relations are expressed by the β-coefficient. The speed with which variables are drawn back to equilibrium, or drift further away from it after a shock in the system, is expressed by the α-coefficient. When a variable's α- and β-coefficient have opposite signs, it means that the variable adjusts towards equilibrium. However, if the α- and β-coefficients show the same sign, the variable drifts further away from equilibrium. If a variable's speed of adjustment is 1, the variable adjusts fully to equilibrium after one time period, which in this study is one year. If a variable's speed of adjustment is less than 1, the variable does not fully adjust to equilibrium within one time period. Finally, if a variable's speed of adjustment is above 1, the variable 'over-adjusts', passing through equilibrium to the other side.

Labour-intensive industry

Cointegration relation (CR1) expresses learning by doing. Since there are no exogenous variables in labour-intensive industry, some undefined external shock pushes the variables away from the steady state relation. TFP adjusts to equilibrium with a speed of 0.29. Machinery investment has a rapid rate of adjustment (2.74), which implies that it 'over-adjusts' and passes through equilibrium to the other side by 1.74. The share of white-collar workers adjusts with a rather slow speed of 0.16. The electricity share drifts further away from equilibrium with a speed of 0.49. (See Table 7.7.)

CR2 expresses negative sector shocks. TFP adjusts back to equilibrium with a speed of 0.30. The speed of adjustment in machinery investments is non-significant and the share of white-collar workers adjusts very slowly, with a

Table 7.7 Identifying cointegration relations and speed of adjustment in labour-intensive, capital-intensive and knowledge-intensive industries

Cointegration relation (CR)	Y	M	E	W	N	t
Labour-intensive industry (χ^2 *df1 p-value 0.97*)						
(H2) CR1 (β-coefficients)	1	−0.12		0.12	0.04	0
Speed of adjustment to CR1 (α)	−0.29*	2.74*		−0.16*	0.49[†]	
(H6) CR2 (β-coefficients)	1	0.26		−1.26	0.61	−0.01
Speed of adjustment to CR2 (α)	−0.30*	0		0.07*	0.10[†]	
Capital-intensive industry model 1 (χ^2 *df2 p-value 0.94*)						
(H1) CR1 (β-coefficients)	1	−1	−0.49	−4.29[‡]	2.70	0
Speed of adjustment to CR1 (α)	−0.12*	0.25*			−0.11*	
(H2) CR2 (β-coefficients)	1	0.11	0.21	−0.11[‡]	0	−0.01
Speed of adjustment to CR2 (α)	−0.86*	0			0.27[†]	
Capital-intensive industry model 2 (χ^2 *df1 p-value 0.62*)						
(H3) CR1 (β-coefficients)	1	−0.68	−0.32	−3.13[‡]	1.90	0
Speed of adjustment to CR1 (α)	−0.13*	0.33[†]			−0.17*	
(H7) CR2 (β-coefficients)	1	0.07	0.19	−0.26[‡]	0.09	−0.001
Speed of adjustment to CR2 (α)	−0.86*	0			0.34[†]	
Knowledge-intensive industry (χ^2 *df2 p-value 0.91*)						
(H5) CR1 (β-coefficients)	1	−0.49	−0.57	−0.51[‡]	0[‡]	0.04
Speed of adjustment to CR1 (α)	0.24[†]	1.29*	0			
(H8) CR2 (β-coefficients)	1	−0.03	0.03	−0.37[‡]	0.13[‡]	0
Speed of adjustment to CR2 (α)	−1.09*	0	−0.42*			

Y = TFP; M = Machinery investments; E = export; W = the share of white-collar workers;
N = the electricity share of total energy; t = trend. 0 = nonsignificant
[‡] exogenous variable
Export is excluded in labour-intensive industry, df = degrees of freedom
* the variable is adjusting back to equilibrium
[†] the variable drifts further away from equilibrium.

speed of 0.07. The electricity share drifts further away from equilibrium with a speed of 0.10. TFP is the only variable that is largely affected by negative sector shocks. Since export is excluded from labour-intensive industry the electricity share alone reflects demand, the fact that the variable moves further away from equilibrium in both cointegration relations after a shock suggest that the undefined factor forcing the system is strongly connected to demand.

Capital-intensive industry model (1) and (2)

Since export and the share of white-collar workers are exogenous, shocks enter the system through these variables. CR1 expresses economies of scale. TFP and electricity adjust rather slowly to equilibrium with a speed of $(-0.12)-0.13$ and $(-0.11)-0.17$ respectively. Machinery investments adjust more rapidly with a speed of $(0.25)\ 0.33$. CR2 expresses diseconomies of scale. TFP adjusts very quickly with a speed of $(-0.86)-0.86$. There is no significant adjustment for machinery investments. Electricity drifts further away from equilibrium with a speed of $(0.27)\ 0.34$. According to neo-classical theory, there is decreasing return to production factors over the long term, suggesting that TFP decreases. The fact that TFP only slowly adjusts to CR1 indicates that economies of scale counteract decreasing factor return, which keeps TFP at a constant level. On the other hand, TFP adjusts rapidly to CR2, which indicates that diseconomies of scale cause decreasing factor return that produces fluctuations in TFP.

Knowledge-intensive industry

CR1 expresses complementarity between subjective knowledge and capital. Since there is no significant α-coefficient for export, and the share of white-collar workers is exogenous, it is probable that shocks enter the system through these variables. TFP drifts further away from equilibrium with a speed of 0.24, whereas machinery investments over-adjust with a speed of 1.29. The process in CR1 may reflect the elasticity of demand for output in knowledge-intensive industry. Because output is both consumed on the goods market and used as a capital good within manufacturing production, it has tremendous potential to affect factor return. CR2 expresses reverse complementarity due to path dependency. TFP adjusts very quickly with a speed of 1.09 and thus 'over-adjusts' to the other side of equilibrium by 0.09. Export adjusts with a speed of 0.40. Machinery investment has no significant adjustment coefficient, which implies that the variable may constitute a driving force behind the process in CR2. The fact that machinery investment continues to increase although demand has decreased reflects a lock-in effect that affects TFP dramatically through rapidly diminishing factor returns. Endogenous growth models assume constant or increasing returns to scale. CR1, reflecting complementarity between subjective knowledge and capital, not only causes constant factor return but also causes increasing factor return, shown by TFP's α-coefficient that drifts further away from the steady state relation. However, CR2 indicates that complementarity processes between human and physical capital also imply lock-in effects and decreasing return to scale due to path-dependency. This is shown by TFP's α-coefficient, which very quickly adjusts to the steady state relation. In the long run path dependency and lock-in effects lead to decreasing returus to capital.

The neo-classical theory of decreasing return to production factors is thus confirmed for knowledge-intensive industries. The reason that TFP does not drift further away from equilibrium in CR1 at the same speed at which it adjusts towards equilibrium in CR2 is that the greater part of knowledge spillover is used to counteract decreasing returns to scale. In contrast, reversed complementarity reinforces decreasing factor return and thus causes huge variations in TFP.

The relationship between machinery investment and TFP

The fact that machinery investments, especially in capital-intensive and knowledge-intensive industries very quickly adjust back to the steady state relation in the first cointegration relation implies that CR1 for each industry strongly affects machinery investments. It is also clear that CR2 for each industry is the most important determinant for TFP variations. Already Adam Smith claimed that capital and output grow together, after him streams of economists have shown a close relation between capital investments and economic growth. This study recognizes two different steady state relations as the most important factors determining variations in machinery investments and TFP. Furthermore it clearly shows that machinery investment is significant in the steady state relation determining TFP for all industries, and that TFP is significant with a large coefficient in the steady state relation determining machinery investments in all industries. Accordingly, even though TFP and machinery investments grow together in the long run, different steady state relations determine their growth paths. Since TFP and machinery investments are cointegrated in both steady state relations reflecting different economic growth processes, the two variables are permanently influencing each other's growth paths.

Identification of the long-run and short-run structure

The cointegrated VAR model has so far been a reduced form model in the sense that the short-run dynamics or contemporary effects are not yet explicitly modelled, but are left in the residual's covariance matrices. Since cointegration relations are able to incorporate long-run relationships between variables, they are of considerable practical importance in econometric modelling. Accordingly, the steady state relations are transformed into variables presented in levels. In the short-run process, we are interested in the direct effect of yearly changes, and individual variables are transformed into first differences. In order to determine the effects of both steady state relations and individual variables, we use a regression model in which TFP constitutes the only dependent variable. The results are shown in Table 7.8. Except for the constant that always is included, explanatory variables are added stepwise so that the variable with the largest R^2 comes first. Thereafter, the two variables with the jointly largest R^2 are added, and so on until all significant variables have been added. We should be aware that the size of the coefficients changes as more variables are added to the model. The final model appears in bold text.

Table 7.8 Explanatory models for TFP in labour-intensive, capital-intensive and knowledge-intensive industries including cointegration relations (CR) and direct effects from other variables (variables are added stepwise after highest R^2)

Labour-intensive industry

C	(H2) CR 1 (−1 lag)	(H6) CR 2 (−1 lag)	(d)Y	(d)M	(d)E	(d)W	(d)N	Correlation coefficient R^2
0.003		−0.27						0.67
0.004	−0.30	−0.36						0.77
−0.01	**−0.59**	**−0.35**					**0.48**	**0.86**

Capital-intensive industry model 1

C	(H1) CR 1 (−1 lag)	(H2) CR 2 (−1 lag)	(d)Y	(d)M	(d)E	(d)W	(d)N	Correlation coefficient R^2
−0.00		−0.88						0.72
−0.02		−1.07					0.66	0.87
−0.02	−0.06	−1.02					0.51	0.89
−0.02	**−0.06**	**−0.99**			**0.16**		**0.50**	**0.90**

Capital-intensive industry model 2

C	(H3) CR 1 (−1 lag)	(H7) CR 2 (−1 lag)	(d)Y	(d)M	(d)E	(d)W	(d)N	Correlation coefficient R^2
0.00		−0.97						0.77
0.02		**−1.09**					**0.57**	**0.88**

Knowledge-intensive industry

C	(H5) CR 1 (−1 lag)	(H8) CR 2 (−1 lag)	(d)Y	(d)M	(d)E	(d)W	(d)N	Correlation Coefficient R^2
0.000		−0.70						0.76
0.01		−0.77					0.33	0.83
0.01		−0.74		0.08			0.26	0.86
−0.01	**−0.21**	**−0.81**		**0.13**			**0.22**	**0.89**

C = Constant; Y = TFP; M = machinery investments; E = export; W = the share of white-collar workers; N = electricity share of total energy. All variables except the CRs are in first difference (d). The CRs have 1 lag, whereas the direct effect is measured for all other variables. Empty cells in the final models imply no significance.

Important findings in labour-intensive industry

The coefficient for CR2 expressing negative structural shocks measures 0.27, which means that when CR2 changes 1, TFP changes by 0.27. CR2 explains 67 per cent of the variations in TFP. Together CR1 and CR2, with the coefficients −0.30 and −0.36 respectively, explain 77 per cent of the variations in TFP. If electricity is added, the model explains 86 per cent of TFP variations. We see that CR1 and CR2 affect TFP with almost equally large coefficients, but that negative structural shocks are the most important factor in explaining TFP variations. This indicates that CR1, expressing knowledge spillovers due to learning-by-doing, neutralizes decreasing factor return and maintains TFP on a constant level, whereas structural changes cause actual variations in TFP.

Important findings in capital-intensive industry model 1

The coefficient for CR2 expressing diseconomies of scale is rather large (0.88), and explains 72 per cent of the variation in TFP. When adding the electricity share (0.66) the two variables explain 87 per cent of TFP variations. CR1 and export are also significant in this model; however, they do not have much impact on TFP variations, as R^2 only increases to 0.90 when added to the model.

Important findings in capital-intensive industry model 2

The coefficient for CR2 expressing diseconomies of scale is very large (0.97) and explains 77 per cent of the variation in TFP. Accordingly, when CR2 changes by 1, TFP changes by 0.97, which shows that diseconomies of scale are a substantial factor and explain a substantial part of TFP variations. The electricity share is the only other significant coefficient (0.57); together the two variables explain 88 per cent of the variations in TFP. Since CR1 expressing economies of scale is not significant, it indicates that economies of scale even out decreasing factor return and maintain TFP on a constant level, whereas diseconomies of scale cause variations in TFP.

Important findings in knowledge-intensive industry

In the model for TFP the coefficient for CR2 expressing reverse complementarity between subjective knowledge and capital due to path dependency is rather large (−0.64), and explains 75 per cent of the variation in TFP. When adding the electricity share, in which the coefficient is 0.33, the model explains 83 per cent. The coefficient for machinery investment is significant, but rather small (0.08). Together the three variables explain 86 per cent of TFP variations. CR1 is the last added variable in the model for TFP, and its coefficient measures −0.21. When all four significant variables are included in the model, it explains 89 per cent of the variations in TFP. The fact that CR1 expressing complementarity between subjective knowledge and capital has a very small effect on TFP indicates that knowledge spillovers due to complementarity between human and physical capital neutralize decreasing factor return, and maintain TFP on a constant level, whereas reverse complementarity causes variations in TFP.

Actual and fitted TFP

Figures 7.1, 7.2 and 7.3 show actual TFP in first difference (full line) and fitted TFP by the final models from Table 7.8 (dotted line). R^2 expresses the correlation between actual and fitted, in this case the percentage of the variations in TFP that is explained by the fitted model. In labour-intensive industry, the model explains 86 per cent of the variations in TFP. In capital-intensive industry, model 1 explains 90 per cent and model 2 explains 88 per cent of TFP variations. In knowledge- intensive industry, the model accounts for 89 per cent of the variations in TFP.

This study shows that steady state relations reflecting 'learning by doing' and that 'negative sector shocks' determine TFP in labour-intensive industry. Perfect competition implies that knowledge is a public good that spills over

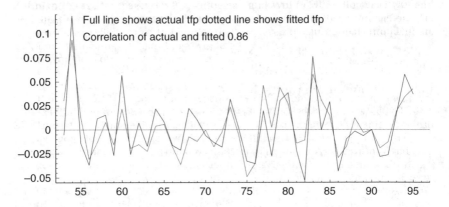

Figure 7.1   Actual and fitted TFP in labour-intensive industry

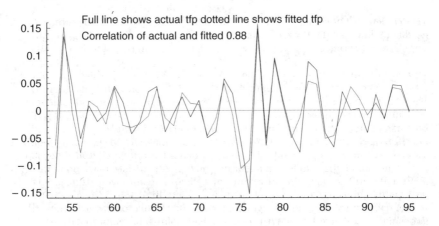

Figure 7.2   Actual and fitted TFP in capital-intensive industry

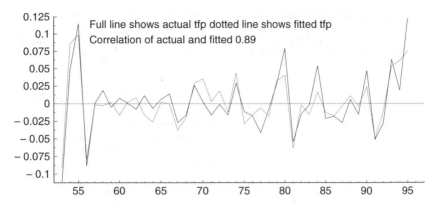

Figure 7.3 Actual and fitted TFP in knowledge-intensive industry

the whole economy so that any industrial production unit has immediate access to all knowledge at zero cost. However, learning-by-doing implies that the labourer learn through experience, and that experience is obtained during the production process. Since it takes some time before such knowledge can spill over to competing firms, competition becomes imperfect. Therefore learning by doing is able to neutralize decreasing factor return. Negative sector shocks imply that if different technologies in diverse sectors produce divergent factor return, relative demand shocks will cause employment variations between those sectors. The idea is that reallocation of labour across sectors is time-consuming. Therefore, employment falls more rapidly in the sectors suffering negative shocks than it rises in the sectors facing favourable shocks due to technological change. Negative sector shocks are not consistent with perfect competition where labour markets would clear immediately after any changes in prices and quantities. Thus negative sector shocks give rise to variations in TFP computed under the assumption of perfect competition.

In capital-intensive industry, steady state relations reflecting economies of scale and diseconomies of scale determine TFP. Neither increasing nor decreasing returns to scale are compatible with the assumption of perfect competition. Sraffa (1926) first demonstrated this fact. Increasing returns to scale neutralize decreasing factor return, whereas decreasing returns to scale reinforce decreasing factor return and thus give rise to variations in TFP computed under the assumption of constant returns to scale.

Finally, in knowledge-intensive industry, steady state relations reflecting complementarity between subjective knowledge and capital, and reverse complementarity due to path dependency and lock-in effects, determine TFP. Complementarity between subjective knowledge and capital generates objective knowledge, which has a large side-production of extended subjective knowledge. It is here assumed that TFP depends on the level of subjective knowledge within a firm compared to other competing firms, assuming that such knowledge is used actively within the production process. Since firms have incentives to protect the secrecy of their discoveries and pro- ductivity improvements, such knowledge would leak out only gradually, and innovating industries would retain competitive advantages for some time. It is thus probable that complementarity between subjective know- ledge and capital neutralizes decreasing factor returns, which is not consist- ent with perfect competition. Moreover, the advantage rapidly increases in the initial phase since the production process gives rise to larger knowledge spillovers. Path dependency implies that a one-chance experimentation with technology encourages additional experimentation with that technology, which increases its advantages over untested alternatives. However it also implies increasing risks due to the loss of return to alternative knowledge investments. Due to the lack of full information, complementarity between subjective knowledge and capital implies lock-in effects, which are not con- sistent with perfect competition. Consequently, reversed complementarity processes due to path dependency cause variations in TFP computed under the assumptions of perfect competition in knowledge-intensive industries.

The results in this study clearly support the alternative view; that TFP variations are caused by output movements arising from sources other than shifts in technology, in contrast to the 'real-business-cycle theory' in which technological shocks constitute the principal mechanism generating TFP. Because output movements across broadly defined sectors have been shown empirically to move in tandem, leading spokesmen for the real business cycle theory suggest the possibility of a unified explanation of business cycles, 'grounded in the general laws governing market economies' (Lucas 1977). The results of this study show that violations of perfect competition and constant return to scale determine 86–90 per cent of the variations in TFP in the three sectors. However, the mechanisms through which perfect competition and constant return to scale are violated differs between sectors. Thus a unified explanation for TFP variations does not appear to exist.

Are the steady state relations identified in this study consistent with the Walrasian equilibrium approach?

Real-business-cycle theory is consistent with the neo-classical Walrasian equilibrium approach, whereas the alternative view argues that equilibrium is unattainable and that variations in output are the result of forces not explained by neo-classical economic theory, such as suggested by Lucas (1977). Equilibrium is not consistent with time series that drift further apart. Since cointegration captures the idea that time series stay together in the long run, cointegration analysis is especially useful for identifying steady state relations. Not all steady state relations are consistent with the equilib- rium approach, however. This study shows that for each industrial sector there are two directions in which the process can be made stationary and thus constitute cointegration relations, expressed as steady state relations between variables. Neither of the identified CRs is consistent with perfect competition or constant return to scale, which are necessary conditions to obtain the unique Walrasian equilibrium. For each sector, CR1 reflects a steady state relation between variables in which factor return exceeds factor cost, and thus neutralizes the law of decreasing factor returns and even gives rise to increasing factor return. CR2 in all sectors reflects a steady state relation in which factor return is below factor cost, which gives rise to dramatic variations in TFP. It is thus the interplay between two conflicting processes which (a) ensures that TFP does not drift further away from equilibrium in the long run, and (b) gives rise to variations in TFP.

The mechanism forcing the economy back to equilibrium in Arrow and Debreu's theory is the equimarginal rule, implying that marginal productivity always equals marginal cost. However, the existence of the equimarginal rule would imply only one steady state relation where factor return equals factor cost. In contrast, this study shows that two steady state relations with opposite impact on factor return, both of which violate the equimarginal rule, govern economic behaviour and determine deviations from equilibrium. Yet the impacts from the two steady state relations are never equally strong, which means that TFP is either below or above equilibrium but never at equilibrium. Over the long term, the processes neutralize each other so that TFP does not develop closer to or further away from equilibrium. In this sense the determinants for TFP variations re-establish the unique Walrasian equilibrium in the long run. Nevertheless, the two steady state relations ruling economic dynamics are characterized by imperfect competition and non-constant return to scale, and not by the equimarginal rule.

Conclusions

Statistical analysis within the growth context frequently treats output as the only dependent variable explained by a number of independent variables in an *economically* well-specified model, and then applies statistical methods to obtain support for their a priori assumptions. This study offers a different way of analysing variations in TFP. A statistically well-defined model is here used actively as a means of analysing the underlying generating process of the phenomena of interest.

The principal aim of this study is to analyse the determinants for variations in total factor productivity in three different types of industries: labour-intensive, capital-intensive and knowledge-intensive industries. The cointegrated VAR model, which is used here, is a dynamic model able to distinguish between endogenous and exogenous variables in a full equation system in which everything is allowed to change; hence there are no *ceteris paribus* or a priori assumptions. Furthermore, the cointegrated VAR model is able to (a) distinguish between short-term and long-term processes in the underlying structure; (b) describe the interactions and feedback effects within a system; and (c) describe the underlying driving forces.

'Real-business-cycle theory' encapsulates the leading explanation for short-term variations in TFP in which technological shocks constitute the key element. The most important macroeconomic evidence for the presence of such technological shocks are the short-term variations in the Solow residual. Therefore TFP is here calculated according to a traditional Solow Cobb–Douglas production function in which the coefficients for labour and capital are assumed to correspond to their marginal effect on TFP. This implies that, if perfect competition and constant return to scale prevail, TFP expresses shifts in technology (Solow 1957). An alternative view is that TFP variations are caused by output movements arising from sources other than shifts in technology: that is, increasing or decreasing returns to scale would cause a Solow residual computed under the assumptions of constant return to scale. In order to determine the degree of variation in TFP that is due to violations of perfect competition and constant return to scale, eight different hypotheses are tested. Each hypothesis corresponds to a steady state relation between variables that violates, in various ways, the assumptions made when calculating TFP.

The results reveal that between 86 and 90 per cent of TFP variations in each type of industry can be explained by violations of perfect competition and constant returns to scale. Moreover, the mechanisms generating such violations are completely different among industries with different methods of production. This study shows that steady state relations reflecting 'learning by doing' and 'negative sector shocks' determine TFP in labour-intensive industry. In capital-intensive industry, steady state relations reflecting economies of scale and diseconomies of scale determine TFP. Finally, in knowledge- intensive industry, steady state relations reflecting complementarity between subjective-knowledge and capital, and reverse complementarity due to path dependency, determine TFP. Accordingly, the results in this study support the alternative theory outlined above, that TFP variations are caused by output movements arising from sources other than shifts in technology, in contrast to the 'real-business-cycle theory'.

It is important to emphasize, however, that these findings do not overlook the importance of technological change in the long term. New technology gives rise to increased capacity for machinery equipment. However, it is the degree to which this capacity is utilized and not technological shocks that, due to production factors' indivisibility, result in increasing or decreasing returns to scale, which in turn affects TFP. Exclusive knowledge gained by the employees gives a firm competitive advantages over other firms until the knowledge is spread to all firms and thus reflects the average level of human capital within the industry. The point is, until exclusive knowledge is spread to all firms the market is imperfect which implies increasing return to scale for the ones using exclusive knowledge. On the other hand, since information is never fully complete, path dependency forces industries to continue investing in certain technologies, despite the fact that both human and physical capital would produce higher returns if utilized in other contexts. Thus, exclusive knowledge and path dependency violate the assumption of perfect competition. In the long run such processes are, of course, dependent on the development of new technologies. However, the short-run variations in TFP are not due to technological shocks but to the exclusivity of the knowledge and variations in external network effects.

Since output movements across broadly defined sectors have been shown empirically to move together, leading spokesmen for the real-business-cycle theory suggest the possibility of a unified explanation of business cycles, grounded in the general laws governing market economies. This study has shown that violations of perfect competition and constant return to scale explain the greater part of TFP variations in all sectors, although the mechanisms through which such violations arise differ among sectors. Another key conclusion drawn from the results in this study is that returns to human capital due to knowledge spillovers are neither constant nor increasing in the long run, as suggested in endogenous growth models, but actually decreasing due to path dependency and changes in external network effects.

Whereas the 'real-business-cycle theory' is consistent with the Arrow–Debreu equilibrium approach, the alternative view argues that equilibrium is unattainable. Cointegration captures the idea that non-stationary time series are kept together in the long run. This study shows that for each industrial sector there are two directions in which linear combinations of variables can be made stationary and thus constitute a steady state relation. Neither of the identified CRs is consistent with perfect competition or constant return to scale, both of which are necessary conditions to obtain equilibrium. For each sector, CR1 reflects a steady state relation between variables in which factor return exceeds factor cost, whereas CR2 reflects a steady state relation in which factor return is below factor cost. To confirm the existence of the equimarginal rule through which the Walrasian equilibrium is obtained, there should be only one identified steady state relation in which factor return equals factor cost, which is not the case here. Accordingly, this study shows that two different processes with conflicting effects on factor return, both violating the equimarginal rule, force economic behaviour and determine TFP deviation from Walrasian equilibrium. In the long run, however, the processes appear to neutralize each other, so that TFP does not develop closer or further away from equilibrium. The determinants for TFP variations thus re-establish Walrasian equilibrium, yet the mechanisms behind that process are inconsistent with neo-classical equilibrium theory, since the equimarginal rule is violated. Consequently, the tools used by Walrasian equilibrium economists are not well suited to analysing the disequilibrium processes by which new technologies are generated and absorbed into the economic structure.

References

Alchian, A. (1963), 'Reliability of progress curves in airframe production', *Econometrica* 31, pp. 679–93.

Arrow, K.J. (1962), 'The economic implications of learning by doing', Review of *Economic Studies*, 29, pp. 155–73.

Arrow, K.J. (1979), 'The division of labour in the economy, the polity and society', *Adam Smith and Modern Political Economy*, ed. G.P. O'Driscoll, Jr (Ames: Iowa State University Press).

Arrow, K.J. and F.H. Hahn (1971), *General Competitive Analysis* (San Francisco CA: Holden-Day).

Arthur, W.B. (1989), 'Competing technologies, increasing returns, and lock-in by historical events', *Economic Journal*, 99, pp. 116–31.

Arthur, W.B. (1990), 'Positive feedbacks in the economy', *Scientific American*, 262, pp. 92–9.

Azariadis, C. and A. Drazen (1990), 'Threshold Externalities in Economic Development', *Quarterly Journal of Economics*, 105 (May), pp. 501–26.

Bengtsson, H. *Ålder och ekonomisk omvandling, – Sveriges tillväxthistoria I ett demografiskt perspektiv, 1890–1995*. Lund studies in Economic History 26

Bernanke, B.S. and M.L. Parkinson (1991), 'Procyclical Labour Productivity and Competing Theories of the Business Cycle: Some Evidence from the Inter-war U.S. Manufacturing Industries', *Journal of Political Economy*, 99 (June), pp. 439–59.

Black, F. (1982), 'General Equilibrium and Business Cycles', *National Bureau of Economic Research* Working paper No. 950 (August).

Calabresi, G. (1968), 'Transaction costs, resource allocation and liability rules': A comment. *Journal of Law and Economics*, 11, pp. 67–73.

Chandler, A.D., Jr. (1962), *Strategy and Structure* (Cambridge, Mass.: MIT Press).

Coase, R.H. (1964), 'The regulated industries': discussion, *American Economic Review*, Paper and Proceedings, 54, pp. 194–7.

Dahlman, C. (1979), 'The problem of externality', *Journal of Law and Economics*, 22, pp. 141–62.

David, P.A. (1985), 'Clio and the economics of QWERTY', *American Economic Review, Paper and Proceedings*, 75, pp. 332–7.

Demsetz, H. (1969), 'Information and efficiency: another viewpoint', *Journal of Law and Economics*, 12, pp. 1–22.

Denison, E.F. (1962), *The Sources of Economic Growth in the United States* (New York: Committee for Economic development).

Denison, E.F. (1967), *Why growth rates differ* (Washington, DC: The Brookings Institution).

Enders, W. (1995), *Applied Econometric Time Series* (New York: John Wiley & Sons).

Engle, R.E. and C.W.J. Granger (1987), 'Cointegration and error- correction: representation, estimation, and testing', *Econometrica*, 55 (March 1987), pp. 251–76.

Goldin, C.D. and Katz, L.F. (1998), 'The origins of technology- skill complementarity', *Quarterly Journal of Economics*, 113, pp. 683–732.

Hansen, H. and K. Juselius (1995), *CATS in RATS: Cointegration Analysis of Time Series* (Estima, Evanston, Illinois).

Hansen, G.D. and R. Wright (1992), 'The labour market in real business Cycle theory', *Federal Reserve Bank of Minneapolis Quarterly Review*, 16 (Spring), pp. 2–12.

Hendry, D.F. (1993), *Econometrics: Alchemy or Science?* (Oxford: Blackwell Publishers).

Hendry, D.F. and G.E. Mizon (1993) 'Evaluating dynamic econometric models by encompassing the VAR', in P.C.B. Phillips (ed.), *Models, Methods and Applications of Econometrics*, (Oxford: Basil Blackwell), pp. 272–300.

Hirsch, W.Z. (1956), 'Firm progress ratios', *Econometrica*, 24(2), pp. 136–44.

Hirschhorn, L. (1984), *Beyond Mechanization: Work and Technology in a Post-industrial Age* (Cambridge, MA:).

Hollander, S. (1965), *The Source of Increased Efficiency: A Study of Dupont Rayon Plants* (Cambridge, MA: MIT Press).

Johansen, S. (1991). Estimations and hypothesis testing of cointegration vectors in Gaussian vector autoregressive models. *Econometrica*, 59, 1551–1580.

Johansen, S. (1988), 'Statistical Analysis of Cointegration Vectors', *Journal of Economic Dynamics and Control*, 12(2/3), pp. 231–54.

Johansen, S. (1996), *Likelihood-Based Inference in Cointegrated Vector Autoregressive Models*, 2nd edn (Oxford: Oxford, University Press,).

Johansen, S. and K. Juselius (1990), 'Maximum Likelihood Estimation and Inference on Cointegration. With Applications to the Demand for Money', *Oxford Bulletin of Economics and Statistics*, 52(2), pp. 169–210.

Johansen, S. and K. Juselius (1992), 'Testing Structural Hypothesis in a Multivariate Cointegration Analysis of the PPP and the UIP for UK', *Journal of Econometrics*, 53(1–3), pp. 211–44.

Johansen, S. and K. Juselius (2000), 'Macroeconomic behaviour, European integration and cointegration analysis', lecture notes.

Jorgenson, D. and Z. Griliches (1967), 'The explanation of productivity change', *Review of Economic studies*, 34, July, pp. 249–83.

Juselius, K. (2003), 'The cointegrated VAR model: econometric methodology and macroeconomic applications', Manuscript, unpublished.

Katz, M. and C. Shapiro (1986), 'Technology and adoption in the presence of network externalities', *Journal of Political Economy*, 94, pp. 822–41.

Kaldor, N. (1966), *Causes of the Slow Rate of Economic Growth in the United Kingdom* (Cambridge: Cambridge University Press).

Kendrick, J. (1973) *Post-war Productivity Trends in the United States, 1948–1969* (New York: NBER) (cumlia University Press).

King, R.G. and Plosser, C.I. (1984), 'Money, credit and prices in a real business cycle', *American Economic Review* 74, June: 363–80.

Koopman, T.C. (1957), *Three Essays on the State of Economic Science* (New York: McGraw-Hill).

Kydland, F. and Prescott, E.C. (1982) 'Time to build and aggregate fluctuations', *Econometrica* 50, November: 1345– 70.

Lieberman, D. (1984), 'The learning curve and prising in the chemical processing industries', *Rand Journal of Economics*, 15, pp. 213–28.

Liebowitz, S.J. and S.E. Margolis (1990), 'The fable of the Keys,' *Journal of Law and Economics* 33:1–25.

Liebowitz, S.J. and S.E. Margolis (1994), 'Network externality: an uncommon tragedy', *Journal of Economic Perspectives*, 8, pp. 133–50.

Liebowitz, S.J. and S.E. Margolis (1995), 'Path dependence, lock in and history', *Journal of Law, Economics and Organization*, 11, pp. 205–26.

Liebowitz, S.J. and S.E. Margolis (1996), 'Should technology choice be a concern for antitrust?', *Harvard journal of Law and Technology*, 9, pp. 283–318.

Long, J.B. and Plosser, C.I. (1983), 'Real business cycles', *Journal of Political Economy* 91, February: 39–69.

180   *Variations in TFP: Sweden*

Lucas, R.E., Jr (1975), 'An equilibrium model of business cycle', *Journal of Political Economy*, 83 (December), pp. 1113–44.
Lucas, R.E., Jr (1977), 'Stabilization of the Domestic and International Economy', *Carnegie-Rochester Series on Public policy*, Vol. 5, ed. Karl Brunner and Allan H. Meltzer (Amsterdam: North-Holland), pp. 7–29.
Lucas, R.E., Jr (1988), 'On the mechanics of economic development', *Journal of Monetary Economics*, 22 (July), pp. 3–42.
Lucas, R.E., Jr and E. Prescott (1974), 'Equilibrium search and employment', *Journal of Economic Theory*, 7 (February), pp. 188–209.
Lundh, C. (1991), *Åldersstruktur och industriell omvandling I Sverige 1945–1985* Lund: Befolkningsekonomiska stiftelsen BEST.
Lundh, C., R. Ohlsson (1994), *Från arbetskraftimport till flyktinginvandring* (Stockholm: SNS (Studieförbundet näringsliv och samhälle).
Maddison, A. (1989), *The World Economy in the Twentieth Century* (Paris: OECD).
Maddison, A. (1995), *Monitoring the World Economy. 1820–1992* (Paris:OECD).
Mankiw, N.G.,D. Romer and D.N. Weil (1992), 'A contribution to the empirics of economic growth', *Quarterly Journal of Economics*, 107(2), pp. 407–37.
Nuwer, M. (1988), 'From batch to flow: production technology and work force skills in the steel industry, 1880–1920', *Technology and Culture* (printed by the Society for the History of Technology, USA), pp. 808–39.
Pettersson, L. (1983), 'Ingenjörsutbildning och Kapitalbildning 1933–1973', *Skrifter utgivna av Ekonomisk-Historiska Föreningen*, Lund, vol. XXXIX.
Prescott, E.C. (1986), 'Theory ahead of business-cycle measurement', in *Real Business Cycles, Real Exchange Rates and Actual Policies*, ed. K. Brunner and A.H. Meltzer, Carnegie-Rochester Conference Series on Public Policy, 25 (Autumn), pp. 11–44. Amsterdam: North Holland.
Romer, D. (1996), *Advanced Macroeconomics*, McGraw-Hill Advanced Series in Economics.
Romer, P.M. (1986), 'Increasing return and long-run growth', *Journal of Political Economy*, 94(5), October, pp. 1002–37.
Schmookler, J. (1966), *Invention and Economic Growth* (Cambridge, MA: Harvard University Press).
Schön, L. (1990), *Elektricitetens betydelse för svensk industriell utveckling*. Vattenfall, SUD, U(S) 1990/60.
Schön, L. (1998), 'Industrial crises in a model of long cycles: Sweden in an international perspective', in T. Myllyntaus (ed.), *Economic Crises and Restructuring in History* (St Katharinen: Scripta Mercaturae).
Schön, L. (2000), "Electricity, technological change and productivity in Swedish industry, 1890–1990", *European Review of Economic History*, vol 4:2.
Sheshinski, E. (1967), 'Test of the learning-by-doing hypothesis', *Review of Economics and Statistics*, 49(4), pp. 568–78.
Solow, R.M. (1957), 'Technical change and the aggregate production Function', *Review of Economic and Statistics*, Vol. 39, Aug.
SOS Industri (Stockholm: Kommerskollegium 1950–60, Statistiska Centralbyrån 1961–94).
Sraffa, P, (1926), 'The laws of returns under competitive conditions', *Economic Journal*, 36, December, pp. 535–50.
Svennilson, I. (1954), *Growth and Stagnation in the European economy* (Geneva: United Nations Economic Commission for Europe).
Uzawa, H. (1965), 'Optimum Technical Change in an Aggregate Model of Economic Growth', *International Economic Review* 6 (January), pp. 18–31.

Welch, F. (1979), 'Effects of cohort size on earnings: the baby boom babies — financial bust', *Journal of Political Economy*, 87, 1979:6, Part 2, pp. 65–95.

Williamson, O.E. (1993), 'Transaction cost economics and organisation theory', *Journal of Corporate Change*, 2, pp. 107–56.

Wright, T.P. (1936), 'Factors affecting the cost of airplanes', *Journal of Aeronautical Sciences* 3(4), 122–8.

Young, A.A. (1928), 'Increasing returns and economic progress', *Economic Journal*, 38, December, 527–42.

8
Spain's Low Technological Level: An Explanation[1]

José M. Ortiz-Villajos

Introduction

The economic backwardness of Spain throughout the Modern period is a well-studied subject.[2] Economic historians have identified a range of causes of this problem: cultural, historical, political, institutional, educational, economic, technological and entrepreneurial. All are inter-related, making it impossible to provide a single, satisfactory explanation for Spanish underdevelopment. Nevertheless, there is general agreement that technological backwardness is one of the most important explanatory factors, and must be studied with special attention.

Despite the importance of the subject, it is only recently that studies have attempted to measure Spain's technological level over the long term. At the same time, few historical studies have explored the causes of Spain's historic backwardness in the field of technology. The present work contributes to research on both questions, focusing particularly on developments during the nineteenth and twentieth centuries.

To quantify the technological level, we have relied mainly on patent data which might imply that this work is a limited exercise. Patents reflect only partially the technological activities of a country since some methods of innovation are not captured by patent records. In addition, patents have certain inherent methodological problems as economic and technological indicators (Griliches 1990), and must be handled carefully. Inventions and other types of innovation are the result of prior investments, particularly in R&D and in education. This chapter pays attention to these variables, all of which are crucial to understanding the process of technical change and the differences in the innovations and patent activities of countries.

Two kinds of patent data have been used for this research: the historical patent series of roughly 30 countries from 1791 until today; and the patents registered in Spain during seven selected years between 1882 and 1935, which we have used to create a database containing qualitative information for close to 16,000 patents. The R&D data used refer mainly to Spain's public expenditure from 1850 to 1965, although more detailed information has been obtained for recent years.

The work is divided into three parts. We begin with a brief explanation of the process of technical change, and the role that inventions (patents), R&D and other variables play in this process. We then use the aggregate patent historical series to demonstrate the low level of creative output in technology in Spain throughout the Modern period. The second part traces the roots of this problem to the traditionally low Spanish investment in R&D activities,[3] by both public and private institutions. The third part, based on the disaggregate information obtained from the Spanish patent records from 1882 to 1935, shows that Spain not only developed less technology than other countries, but that this technology was applied mainly to the more traditional (technologically less complex) industrial sectors. Although this is a sign of Spain's technological shortcomings, it is also a logical outcome: in Spain, the dominant industrial branches were the traditional ones, so it makes sense that the technology patented was applied to these branches. From this point of view, patents reveal a degree of complementarity between Spanish industrial skills and the development of new technology. Nevertheless, the information obtained from patent records clearly demonstrates Spain's technological backwardness compared to other European countries. Although the cause of this historic gap is not a simple one and needs more study, this chapter offers a possible explanation. A particular set of historical and political circumstances discouraged Spain from investing in R&D and education and, at the same time, contributed to the emergence of a type of Spanish entrepreneur less inclined to innovate than to seek protection. We argue that Spain did not lack the genius or ability to innovate, but simply the kind of 'knowledgeable entrepreneurship' prone to innovation and competition that, for instance, McInnis (in Chapter 3) has described for Canada in the same period.

The technological level of Spain in the international context

Although few would deny that technology has been instrumental in the extraordinary progress of modern times, there has been no systematic effort to gather and analyse science and technology indicators over the long term. Consequently, for much of the Modern period there is little comparable and continuing data on the evolution of the technological levels of countries. Although patents are only a partial indicator of the results of R&D activities, they can be used to obtain an approximate measure of the technical levels of various countries. But how do inventions influence technical progress, and what are their relationships with R&D activities? An answer to this question depends on our understanding of the process of technical change,[4] which is represented in Figure 8.1.

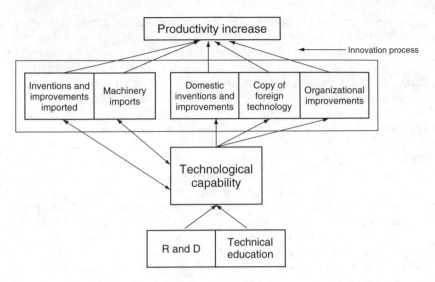

Figure 8.1   The process of technical change

In this process, the final object of which is increased productivity, the central variable is *technological capability:*[5] the accumulated knowledge that allows a company or an economy to produce, copy or adapt technical improvements of every kind (inventions, importation of technology, organizational improvements, and so on). These technical improvements, once implemented in the production process, can be considered as innovations, in that they immediately increase productivity. But technical capability does not arise spontaneously: it is the result of a protracted process of public and private efforts to foster education and research in science and technology. In general, the technical capability of an economy is the result of long-term investments in R&D and technical education. These factors might not influence the innovation process directly or immediately, but without them an economy will not be able to develop its own technology, or copy or adapt foreign technology. The economy that channels resources into these investments will achieve a much higher technological and economic level in the long term.

As Figure 8.1 shows, invention is not the only method of innovation. There are others, including the importation of machinery and technicians, imitation, organizational improvements and so on. By focusing on patents, we are excluding from analysis other modes of innovation that may be even more important in the process of technological development. Despite these limitations,[6] patents are very useful in comparing the technological levels of countries. The patent system is widely used (as proved by the growing use of patents in all countries during the last two centuries), and it records developments in domestic technology as well as imported technology. Indeed, many studies have demonstrated the close relationship between technical education and R&D investment, and patent registration. Thus patents can be considered indications of the output of R&D and technical education investments, and have become one of the most reliable tools of analysts and institutions for measuring the technical levels of countries.

In 1994 Spain obtained 9.6 patents per million inhabitants in the European Patent Office. In the European Union it ranked third from last, ahead of only Greece (4.4 patents per million) and Portugal (1.6 patents per million), and far behind Germany (154.7 patents per million), the highest ranking country.[7] These data reveal the degree of Spanish technological backwardness in relation to most European countries in recent times. But what has been the situation in the past? To answer this question, we have examined the patent statistics for the nineteenth and twentieth centuries.[8] Part of this data is summarized in Figure 8.2, comparing the evolution of the total number of patents (domestic and foreign) per inhabitant applied for in nine different countries between 1820 and 1985.[9] It shows that Spain's technological backwardness, relative to the most advanced countries in Europe during the last two centuries, has been historically consistent. This becomes more evident when we consider only the patents applied for by residents. Domestic patents are an output of internal R&D activities, and can be considered an approximate indicator of domestic technological capability. Figure 8.3 shows

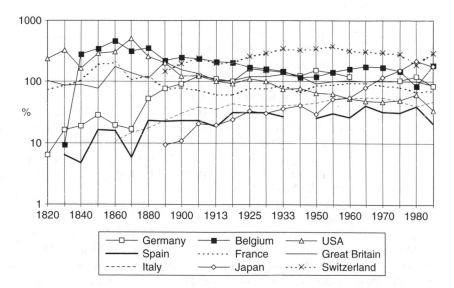

Figure 8.2 Total patent applications per thousand inhabitants in 9 countries: percentage over the average value for 23 countries, 1820–1985 (logarithmic scale)

Source: Ortiz-Villajos (1999), graph 2.10, p. 83.

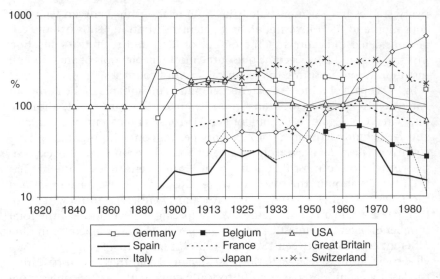

Figure 8.3　Domestic patent applications per thousand inhabitants in 9 countries: percentage over the average value for 23 countries, 1820–1985 (logarithmic scale)

Source: Ortiz-Villajos (1999), graph 2.11, p. 84.

that Spain has lagged behind in this category, providing further evidence that the Spanish R&D effort has been weaker than in other countries.

To better understand the Spanish case, we can compare it with the Japanese one. Until the beginning of the twentieth century, Spain had a similar or even higher technological level than Japan, when measured in terms of the number of patents per inhabitant (Figure 8.2). As is well known, Japan was then making great strides in the fields of science and technology, owing to its investment in R&D. This trend accelerated in the decades after the Second World War when Japan became one of the most technologically advanced countries in the world. This was reflected in the extraordinary increase in the number of patents per inhabitant in Japan from 1950 onwards. In contrast, Spain maintained a ratio of patents per inhabitant that was inferior to most other countries, and over the last 50 years the gap has increased. The available data on the registration of domestic patents in Spain (Figure 8.3) correspond with her reduced competitiveness in most modern sectors and account (at least in part) for her inability to attain the technical level of the leading developed nations. Patent data can be used, then, as an indicator of the technological gap between nations, and even to explain the different economic growth rates of nations, as Fagerberg (1987) has done. We have made a similar analysis (Ortiz-Villajos 1996), and have found a positive and significant correlation between patents per inhabitant and GDP per head for various nations.

Research and development expenses in Spain, 1850–1998

How can we explain the marked technological backwardness of Spain that is revealed in the patent data? A key factor is insufficient investment in R&D. Responsibility for this can be attributed both to the Spanish government and to entrepreneurs: to the governments, because the promotion of science and technology activities and institutions has been traditionally for them a secondary object; and to the entrepreneurs, because they have expended few resources and little effort in promoting technological innovation.[10] These are considered to be the main problems of the present Spanish innovation system.[11]

Table 8.1 indicates that even though the Spanish 'technological effort' (R&D expenses as a percentage of GDP) increased from 0.61 per cent to 0.88 per cent between 1986 and 1998, in its best year it amounted to only 46 per cent of the technological effort of the European Union (1.91), and only 33 per cent of that of the USA (2.64). If we focus only on the R&D expenses implemented and financed by private companies (Table 8.2), the disparity is even greater. Thus Spain not only devotes a smaller percentage of resources to R&D activities, but finances these activities primarily through public institutions, while in the EU and the USA, the main financiers and implementers are private companies. At the same time, the public R&D institutions – the universities, the CSIC[12] and others – have a weak relationship with the private sector, so their technological efforts have little effect on the economy. This is an ongoing problem for the CSIC, the most important Spanish scientific research institution, responsible for 20 per cent of Spanish

Table 8.1 R&D expenses as a percentage of GDP, 1986–98

Year	Spain	EU	USA
1986	0.61	1.95	2.85
1987	0.64	2.00	2.82
1988	0.72	1.99	2.78
1989	0.75	1.99	2.73
1990	0.85	1.99	2.78
1991	0.87	1.98	2.81
1992	0.91	1.96	2.74
1993	0.91	1.98	2.61
1994	0.85	1.94	2.51
1995	0.85	1.92	2.61
1996	0.87	1.90	2.62
1997	0.86	1.91	2.64
1998	0.88	1.91	2.64

Source: Martín (1999), table 1, p. 7.

Table 8.2   Percentage of R&D expenses implemented and financed by companies

Spain Year	%	% financed implemented	EU %	% financed implemented	USA %	% financed implemented
1986	56.5	50.1	66.5	51.9	74.9	51.4
1987	55.9	47.9	66.2	53.1	74.8	50.2
1988	57.6	48.7	66.5	54.6	74.4	51.5
1989	56.9	48.6	66.6	54.5	73.7	53.7
1990	58.5	48.2	66.3	53.8	73.0	55.4
1991	56.5	48.7	64.2	53.2	76.1	59.1
1992	51.1	44.3	64.0	54.3	75.4	60.0
1993	48.7	42.0	63.1	53.9	74.4	60.1
1994	47.7	41.3	62.9	54.2	74.3	60.6
1995	49.4	45.3	63.0	54.2	75.2	62.3
1996	49.7	46.5	63.2	54.8	76.3	64.2

Source: Martín (1999), table 2, p. 8.

scientific production, as Mulet (1998, pp. 52–3) has remarked: 'The different institutions of the Council [CSIC] have not adopted suitable systems to transfer their results to the private sector ... Some of these institutions are making efforts to develop their technology, but not to transfer it, something that must begin with closer contact with the industry' my translation). The fact that the public sector is the principal investor in the Spanish R&D system is a problem, but more serious is the weak relationship between R&D and the private sector. If we keep in mind the small relative size of the Spanish budget, the conclusion is not only that Spain spends much less on R&D, but that the effectiveness of this expense is minimal.

This has been the situation in recent years. What has happened in the past is more difficult to know, owing to a lack of information. Although we have anecdotal evidence, and descriptions of R&D activities in Spain throughout the Modern period,[13] there is little quantitative data. General estimates of R&D expenditures were not made until the 1970s. However, it is possible to approximate the amount of public R&D expenditure by studying Spain's general budget, which was published beginning in 1850. Although the budget did not contain any item specifically concerning R&D activities until 1907, when the *Junta para la Ampliación de Estudios* (JAE)[14] was created, Aracil and Peinado (1976) have made estimates of the public R&D expenditures for the 1850–1965 period. These provide some insight into the evolution of the state effort to promote scientific and technological research in Spain.

As shown in Figures 8.4 and 8.5, after the stagnation observed during the nineteenth century, the level of public expenditure devoted to research activities embarked upon a growth trend in 1910 that continued until the Spanish

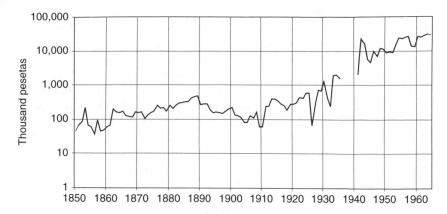

Figure 8.4 Public R&D expenses in Spain, 1850–1964 (thousand pesetas in constant prices of 1958, logarithmic scale)

Sources: Aracil and Peinado (1976); Prados de la Escosura (1993), series PRGDP58.

Figure 8.5 Public R&D expenses in Spain: percentage of total public budget, 1850–1964 (logarithmic scale)

Source: Aracil and Peinado (1976).

Civil War (1936–9). This was due to the creation of the JAE in 1907 and of the *Fundación Nacional para Investigaciones Científicas y Ensayos de Reformas*[15] (FNICER), another public R&D institution, in 1931. The growth spurt in public R&D expenditure after the civil war was due in large part to the creation of the CSIC. In absolute terms (Figure 8.4), this expenditure continued to increase annually, with some variations, until 1965. But the proportion of R&D expenditure in the overall budget (see Figure 8.5) reached its maximum level in 1954 (1.16 per cent). Thereafter it decreased, at least until 1965.

Although R&D public expenses began to be significant in the twentieth century, it is important to note that they had experienced a growth trend many years before, beginning in 1860. This trend ended abruptly in 1889, however, when public expenditure in R&D reached a level (0.053 per cent of the total budget) that was not achieved again until the 1930s. The decline between 1890 and 1907 was significant, and more intensive study of the scientific policy of these years is required in order to understand the Spanish technological gap. Two important historical circumstances appear to be explanatory factors for the R&D expenses decline: the War of Cuba and the increase of protectionism in Spain. The War of Cuba had tremendous consequences for Spanish society from its outbreak in January 1895. The problems that led to the war had commenced five years earlier when the USA – with growing economic and political interests in Cuba – threatened to boycott Cuban sugar, the economic mainstay of the island. Although Spain temporarily reached an agreement with the USA, the American government rejected the treaty in 1893, stimulating an internal revolt against Spain that began in 1895. As a result, military expenses increased significantly during the 1890s, funnelling resources away from other areas, including public expenditure in R&D.

A second probable cause of the decline in R&D was Spain's embrace of protectionism beginning in 1891. From that year, the Spanish government dedicated more laws to protecting and subsidizing its economy. In the long term, the strategy of protectionism short-circuited the R&D dynamism of the 1880s and made chronic the scientific and technological backwardness of Spain.[16] In an excessively protected market Spanish companies did not have the incentive to compete. Moreover, even in the more expansive years of R&D expenditure, its proportion of the overall national budget was minimal. If we consider it in proportion to Spain's GDP, it was virtually insignificant. In 1964 Spanish R&D expenditure (both public and private) amounted to only 0.2 per cent of GDP, far below the figures of more technologically-developed countries. Some regressions have been made between the time series of public R&D expenditure and the evolution of the GDP per head between 1850 and 1965. The results show no significant correlation between the two variables. Although this evidence might be weak, it is consistent with the main problems we have identified with Spanish R&D institutions: their insufficient financial investment and their weak connection with industry. Thus it can be maintained that public science and technology policies have traditionally made a small contribution to Spanish economic growth. In all likelihood, the most important contributions of the state to Spain's technological progress – and to the country's economic development – have been the creation of technical education institutions, such as engineering schools, and various legal measures implemented to protect and encourage both innovation and the import of foreign technology.[17]

Although public promotion and support of science and technology are important for technological progress, the decisive factor is the private R&D effort. One of the main aims of public policy must be to encourage the private innovation process. This is especially important in Spain, where the scarcity of private companies with R&D departments has plagued the National Innovation System throughout the Modern period: 'The lack of a science-based industry could not be addressed by public institutions alone, no matter how right their orientations might be. Scientific progress is a complex problem, and it is not only advanced by public efforts ... but also by private and industrial ones' (Sánchez Ron 1999, p. 169; my translation). In short, the main problem of the Spanish R&D system during the Modern period has been its lack of a scientific and technological entrepreneurial structure, whereas in other countries industry has worked since the nineteenth century to create R&D departments and institutions. Unlike other countries, in which the state has mainly coordinated and supported private efforts, in Spain the state has been the pioneer in the creation of the R&D structure. But this public effort has been limited and unable to stimulate significant scientific and technological growth in Spanish industry and society. In truth, the responsibility for insufficient entrepreneurial investment in R&D is not only the fault of Spanish entrepreneurs, but also of the state, which has not sufficiently stimulated R&D investment, and may even have discouraged it through a system of excessive protectionism.

The technological backwardness of Spanish industry: evidence from patent records, 1882–1935

The historical patent series has shown that throughout the Modern period Spain has maintained a low technological level. We have argued that the responsibility for this can be attributed both to the state and to the private sector, but we have no statistical evidence of private technological investment in the long term, making it difficult to trace its history. Currently, the only long-term homogeneous technological data set available is the one contained in the patent records. As we have argued, patents are a useful, though partial, reflection of the results of R&D investments, although for many reasons this part is a significant one; the usefulness of aggregate patent data is demonstrated in the first part of this chapter. The principal value of patents is the information that can be obtained from each document: the nationality and type of inventor, industrial sector, company, place of residence, relevance of the invention and so on. This type of disaggregate information forms the basis of the current analysis. We have classified all the patents registered in the Spanish Patent Office during seven selected years from 1882 to 1935 (1882, 1887, 1897, 1907, 1917, 1922 and 1935); almost 16,000 patents in all (see Table 8.3).[18] Based on this information, we have identified some general characteristics of the Spanish patent system during those years. First, foreign patents dominated throughout the period,

Table 8.3 Number and percentage of patents applied for in Spain by countries of origin, 1882–1935

Country	1882 Pat.	%	1887 Pat.	%	1897 Pat.	%	1907* Pat.	%	1917 Pat.	%	1922 Pat.	%	1935 Pat.	%	Total Pat.	%
Spain	266	31.2	432	37.4	655	36.0	1,010	39.9	1,598	67.7	1,610	47.2	1,909	53.7	7,480	47.66
France	209	24.5	230	19.9	332	18.3	410	16.2	147	6.2	467	13.7	265	7.4	2,060	13.13
Germany	70	8.2	121	10.5	261	14.3	439	17.3	78	3.3	335	9.8	440	12.4	1,744	11.11
UK	90	10.6	140	12.1	178	9.8	240	9.5	158	6.7	281	8.2	225	6.3	1,312	8.36
USA	125	14.7	121	10.5	131	7.2	112	4.4	169	7.2	244	7.1	171	4.8	1,073	6.84
Italy	8	0.9	14	1.2	29	1.6	97	3.8	33	1.4	108	3.2	102	2.9	391	2.49
Switzerland	4	0.5	10	0.9	11	0.6	12	0.5	52	2.2	99	2.9	103	2.9	291	1.85
Holland	4	0.5	1	0.1	5	0.3	3	0.1	14	0.6	44	1.3	109	3.1	180	1.15
Belgium	22	2.6	22	1.9	36	2.0	11	0.4	4	0.2	49	1.4	34	1.0	178	1.13
Austria	18	2.1	14	1.2	25	1.4	39	1.5	8	0.3	25	0.7	35	1.0	164	1.05
Sweden	2	0.2	6	0.5	14	0.8	40	1.6	29	1.2	35	1.2	29	0.8	155	0.99
11 countries†	818	96.0	1,111	96.1	1,677	92.2	2,413	95.2	2,290	97.0	3,297	96.6	3,422	96.2	15,028	95.76
All countries	852	100	1,156	100	1,819	100	2,534	100	2,362	100	3,413	100	3,558	100	15,694	100

* Estimated data.

† Countries that applied for more than 1 per cent of the patents.

Sources: BOPI (*Industrial Property Official Bulletin*) and Patent Record Books (OEPM).

although Spanish applicants had a growing participation, obtaining roughly 30 per cent of all the patents in 1882 and 54 per cent in 1935. Second, individual applicants were the most numerous patentees, although companies increased their participation from 8.2 per cent of all patents in 1882 to 41.6 per cent in 1935. Finally, a large proportion (68.2 per cent) of entrepreneurial patents – an indication of the modernization of the technological structure of a country – belonged to foreign companies. These data provide early evidence of the technological backwardness of Spanish industry.

During the period 1882–1935, almost 96 per cent of the patents registered in Spain were applied for by patentees from eleven countries. Ranked in order based on their number of patents,[19] they were: Spain, France, Germany, UK, USA, Italy, Holland, Switzerland, Belgium, Austria and Sweden. The classification of inventions by sector of implementation allows us to observe the level of innovative activity in each of the different branches of the economy (Table 8.4). Between 1882 and 1935, inventiveness in Spain (as in many countries during that period) was orientated mainly to the manufacturing sector: close to 90 per cent of patents involved this sector. Consequently, the importance of the four main sectors (agriculture, manufacturing, construction, and services) in the GDP[20] was not directly related to the number of patents applied for in each. However, there was a correlation over the long term: industry and services increased their importance (as measured in the number of patents as well as in their participation in the GDP), while the importance of agriculture and construction declined in both aspects.

If we consider both the sector and nationality of patents, we can observe key differences between foreign and domestic technological activity in Spain. The distribution of patents into branches shows the different sectors targeted by each country and also provides an idea of the technological levels of each country.[21] Table 8.5 summarizes this information, distinguishing between foreign and domestic patents.[22] As an index of the foreign and domestic technological levels in each sector, we have used the Revealed Technological Advantage indicator (RTA): the percentage of one country (or a group of countries) of the patents applied for in one specific sector, divided by the percentage of this country (or a group of countries) of all the patents applied for by all the countries in all sectors.[23] It can be interpreted as follows: if RTA<1, the country (or group of countries) participates in a sector with a lesser proportion of patents than its participation in the total amount, suggesting that it is at a technological disadvantage in that sector; if RTA>1, the country participates in the sector with a greater proportion than in the total and can be said to have a technological advantage in that sector.

Table 8.5 shows that domestic inventors focused their energies on traditional sectors, which were technologically less complex, including machinery, textiles and other industries. Foreigners, on the other hand, applied for patents mainly in machinery, electric and electronic material, and transportation equipment. The figures show that while non-residents

Table 8.4 Patent applications in Spain in four economic sectors, 1882–1935

Sector	1882 No.	%	1887 No.	%	1897 No.	%	1907 No.	%	1917 No.	%	1922 No.	%	1935 No.	%
Agriculture	33	3.9	39	3.4	27	1.5	42	1.7	40	1.7	54	1.6	51	1.5
Manufacturing	753	88.7	1,020	88.5	1,662	91.9	2,203	89.9	2,117	91.5	3,054	91.2	3,230	92.1
Construction	37	4.4	47	4.1	68	3.8	104	4.2	59	2.6	121	3.6	140	4.0
Services	26	3.1	47	4.1	51	2.8	102	4.2	97	4.2	118	3.5	87	2.5
Total	849	100	1,153	100	1,808	100	2,451	100	2,313	100	3,347	100	3,508	100

Sources: BOPI and Patent Record Books (OEPM).

Table 8.5 Domestic and foreign patent applications in Spain by sector of use, and RTA, 1882–1935 (seven selected years)

Sector	Domestic patents				Foreign patents				Total patents	
	Pat.	% (1)	% (2)	RTA	Pat.	% (1)	% (2)	RTA	Pat.	% (1)
Agriculture, animal fishing products,	176	2.4	61.5	1.3	110	1.4	38.5	0.7	286	1.9
Energy and water	282	3.8	34.5	0.7	536	6.6	65.5	1.2	818	5.3
Primary metal manufacturing	104	1.4	33.5	0.7	206	2.5	66.5	1.3	310	2.0
Non-metallic mineral mining	314	4.3	53.4	1.1	274	3.4	46.6	0.9	588	3.8
Chemical industry	527	7.2	44.9	0.9	647	8.0	55.1	1.0	1,174	7.6
Fabricated metal product	351	4.8	51.3	1.1	333	4.1	48.7	0.9	684	4.4
Machinery	1,201	16.4	38.3	0.8	1,928	23.8	61.7	1.2	3,129	20.3
Electric and electronic material	518	7.1	35.2	0.7	953	11.8	64.8	1.2	1,471	9.5
Transportation equipment	464	6.3	36.6	0.8	805	9.9	63.4	1.2	1,269	8.2
Food	428	5.8	59.4	1.3	292	3.6	40.6	0.8	720	4.7
Beverages and tobacco	213	2.9	54.6	1.2	177	2.2	45.4	0.9	390	2.5
Textile industry	934	12.7	67.1	1.4	458	5.7	32.9	0.6	1,392	9.0
Wood and furniture	207	2.8	62.7	1.3	123	1.5	37.3	0.7	330	2.1
Paper and graphic arts	232	3.2	52.0	1.1	214	2.6	48.0	0.9	446	2.9
Other industries	743	10.1	56.4	1.2	575	7.1	43.6	0.8	1,318	8.5
Construction	284	3.9	49.3	1.0	292	3.6	50.7	1.0	576	3.7
Wholesale trade	7	0.1	70.0	1.5	3	0.0	30.0	0.6	10	0.1
Retail trade	47	0.6	59.5	1.3	32	0.4	40.5	0.8	79	0.5
Restaurants and catering	11	0.2	78.6	1.7	3	0.0	21.4	0.4	14	0.1
Transport and communications	22	0.3	44.0	0.9	28	0.3	56.0	1.1	50	0.3
Financial institutions and services to companies	159	2.2	84.1	1.8	30	0.4	15.9	0.3	189	1.2
Other services	104	1.4	55.9	1.2	82	1.0	44.1	0.8	186	1.2
Total	7,328	100	47.5		8,101	100	52.5		15,429	100

(1) Percentage of patents of the sector in each group of applicants
(2) Percentage of patents of each group of applicants in the sector
Sources: BOPI and Patent Record Books (OEPM).

had a technological advantage (RTA>1) in the most modern sectors (energy, basic metals, chemical, machinery, electric material, transportation equipment, and transport and communications), Spanish applicants were at a technological disadvantage in all of these sectors. However, it is important to note that the focus of domestic patentees on traditional sectors is in accordance with the structure of the Spanish economy, which was dominated by these sectors. It can be interpreted as a sign of the rational attitude of domestic inventors and patentees, who directed their skills and efforts towards inventions that were more likely to be in demand. It can be argued that patent data show a certain complementarity between technology and skills in Spanish industry during this period.

Yet this complementarity was less apparent in agriculture if we consider the extremely low proportion of patents directed to this sector in Spain. During the period under discussion, this sector was the most important in the Spanish economy, accounting for more than 50 per cent of the work force. Although it is true that invention was in all countries chiefly an industrial phenomenon, it is also true that the number of patents connected with the agricultural sector in the Spanish Patent Office was especially low. In England, the patents relating to agriculture in 1800 – a sector that employed at that time over one-third of the work force – accounted for roughly 4 per cent of patents (MacLeod 1988, p. 97). In Spain – which had a relatively larger agricultural sector – agriculture accounted for just 1.5 per cent of patents in 1897 (Table 8.4).[24] It seems that in Spanish agriculture there was less complementarity between technology and skills. It is well known that the slow introduction of new farming techniques led to low productivity within the Spanish agricultural industry, and contributed to Spain's economic backwardness during the Modern period.[25]

As we have explained (Tables 8.1 and 8.2), insufficient R&D private investment is currently the weakest area of the Spanish innovation system. Although anecdotal evidence indicates that this problem also existed in earlier times, we lack the R&D statistics to state this with absolute certainty, but patent records can help to solve the problem. Using our database, which can differentiate individual from entrepreneurial patents, it is possible to obtain the information we are looking for, as patents applied for by companies represent an approximate indicator[26] of the degree of entrepreneurial R&D investment. Table 8.6 differentiates individual from entrepreneurial patents in the eleven most important countries that applied for patents in Spain from 1882 to 1935. Spanish patentees accounted for the lowest proportion of entrepreneurial patents (18 per cent) compared with all other countries except for Austria, which was also an economically backward country at that time. This supports our argument for the weakness of private R&D investment in Spain during the period under discussion.

Moreover, the information contained in Table 8.6 fits with what we know was happening internationally during that period: many companies were

Table 8.6 Patents registered in Spain by country and type of applicant, 1882–1935 (seven selected years altogether)

Type of applicant	Spain Pat.	% (1)	% (2)	France Pat.	% (1)	% (2)	Germany Pat.	% (1)	% (2)	UK Pat.	% (1)	% (2)	USA Pat.	% (1)	% (2)
Individuals	6,131	82.0	53.5	1,446	70.2	12.6	972	55.7	8.5	887	67.6	7.7	654	61.0	5.7
Companies	1,348	18.0	31.8	614	29.8	14.5	772	44.3	18.2	425	32.4	10.0	419	39.0	9.9
Total	7,479	100	47.7	2,060	100	13.1	1,744	100	11.1	1,312	100	8.4	1,073	100	6.8

Type of applicant	Italy Pat.	% (1)	% (2)	Switzerland Pat.	% (1)	% (2)	Holland Pat.	% (1)	% (2)	Belgium Pat.	% (1)	% (2)	Austria Pat.	% (1)	% (2)	Sweden Pat.	% (1)	% (2)
Individuals	276	70.6	2.4	131	45.0	1.1	39	21.7	0.3	124	69.7	1.1	143	87.2	1.2	105	67.7	0.9
Companies	115	29.4	2.7	160	55.0	3.8	141	78.3	3.3	54	30.3	1.3	21	12.8	0.5	50	32.3	1.2
Total	391	100	2.5	291	100	1.9	180	100	1.1	178	100	1.1	164	100	1.0	155	100	1.0

(1) Percentage of patents of the category in each country
(2) Percentage of patents of the country in each category
Sources: BOPI and Patent Record Books (OEPM).

creating their own R&D departments, which were emerging as more import-ant inventors than individuals. It is reasonable to deduce that those countries with a greater proportion of entrepreneurial patents were those with greater degrees of private R&D investment. Thus the small proportion of companies among the Spanish patentees reflects the lesser degree of private investment in technical innovation. Overall, the trend towards innovation in more trad-itional sectors (Table 8.5), and the marked shortage of companies among Spanish patentees (Table 8.6), help to explain the absence of great Spanish industrial companies.[27]

Conclusion

In the first section above, we used the international historical series of pat-ents to demonstrate Spain's persistent technological under-development. One of the roots of this problem is the historically low level of R&D invest-ment in Spain, both public and private, as described in the second part. The third section considers the results of a detailed study of the patents registered in the Spanish Patent Office between 1882 and 1935. Analysis of this source has allowed us to define some of the characteristics of the Spanish technolo-gical system. The prevalence of foreign patents in high-technology sectors, and the scarcity of patents applied for by domestic companies, indicate that Spanish entrepreneurs had less technological proficiency than foreigners (a consequence of under-investment in R&D). The key question is why Span-ish entrepreneurs invested so little in technology. Although the answer to this question is not an easy one, the purpose of this study has been to for-mulate some hypotheses, in the hope of fostering discussion and further research.

It has been argued that the peculiar historical situation of Spain in a European context encouraged Spanish governments and institutions to devote less effort to social investments, including investments in education.[28] As a consequence, nineteenth-century Spain exhibited compara-tively low educational[29] and technological levels. It has been argued that this peculiarity contributed to configure a society less prone to innovation, and with a lower rate of innovative entrepreneurs, than other societies.[30] The low proportion of companies among the Spanish patentees between 1882 and 1935 is an indication of this.

For these reasons, the industrial development of Spain was slow and dependent upon foreign capital, technology and technicians. As a con-sequence, Spanish entrepreneurs demanded protection from the state, and successive governments increased this protection from the last quarter of the nineteenth century onwards. This political strategy contributed to the low technological investment of Spanish companies, which preferred to rely on protectionism rather than increase their competitiveness and exports through innovation. The outstanding contribution of a number of well-known Spanish inventors, scientists and engineers shows that Spain has not lacked talent or genius in the technological field. If inventors such as Torres Quevedo, La Cierva or Peral did not become the Spanish Edison, Marconi, Siemens or Phillips, it was due in large part to the long-term configuration of a social and economic climate that discouraged competition and innovation.

These circumstances combined to produce a chronic technological under-development in Spain, especially in high technology sectors. Since this technology was often indispensable for industrial and economic develop-ment, the government permitted the introduction of some technology and technicians, and even the transplantation of foreign companies. So, notwith-standing the high level of protectionism, foreign capital and technology did reach Spain, but not to the extent that was required to foster its own develop-ment. Accustomed to economic protectionism, many Spanish entrepreneurs did not exercise their abilities to innovate, and when the economy began to open in the 1960s they were not prepared to compete with foreign com-panies, particularly in the more sophisticated technology sectors. In other industrial sectors, many Spanish entrepreneurs and companies were able to compete and enjoyed a significant presence in some international markets, especially after the economic liberalization of the 1960s. During those years, Spanish productivity grew rapidly. This expansion took place in tourism and in some traditional industrial sectors (textiles, food, chemistry, shipbuild-ing, and so on), as well as in more modern sectors, such as the automobile industry. Technical innovation was a key factor in this expansion, but it was not the result of domestic R&D investments; rather, it came mainly from abroad, as Cebrián and López (2002) have demonstrated. Foreign technology and capital imports began to grow rapidly after the great liberalization of 1959. Many foreign companies established branches in Spain, forcing many Spanish entrepreneurs to buy patents or to sign agreements of technology transfer with foreign companies.

To make a generalization (acknowledging that there are always excep-tions), it may be said that the Spanish model during the nineteenth and twentieth centuries was to specialize in traditional industrial sectors (with lower R&D and human capital requirements), and to modernize through the import of foreign technology and the establishment of foreign companies in Spain. This model has made Spain a follower rather than a leader in the development of technology. If we examine the economic results, this strategy appears to have worked to a degree. By the mid- 1970s (according to the OECD statistics), Spain had joined the leading group of developed countries and has continued developing during the past few decades. The question is whether this development trend will continue. The persistently high unemployment rate in Spain – the highest in the EU for decades – may be a harbinger. The traditional model has its limits, as the rate of unemploy-ment will be difficult to reduce if the country does not create employment opportunities in more sophisticated technology sectors. Although specialization in traditional sectors may have been sufficient in the past, the situation has changed. In recent years, certain developing countries have become more competitive than Spain in the textiles industry and in other important traditional sectors, including tourism. Under these circumstances, advancing the technological level of Spain through innovation, backed by investment and education, appears not only desirable, but necessary.

Notes

1 This work is based on a paper presented to the XIII International Economic History Congress, Session 32 (Buenos Aires, 22–26 July 2002). I thank Jonas Ljungberg, whose comments on the first version of this chapter were extremely useful. An anonymous commentator made some valuable remarks on the second version. I also thank the participants in the Workshop on Technology and Human Capital in Historical Perspective (Lund, 26–28 September 2003), for their comments and suggestions. Any mistakes and omissions can only be attributed to the author.

2 Many studies have attempted to account for Spanish economic backwardness. It is not possible to mention all of them, but some of the more relevant and recent ones are, for instance: Carreras (1992), Comín and Martín Aceña (1996), Cubel and Palafox (1997), Fraile Balbín (1999), Prados, de la Escosura, Daban and Sanz (1993), Prados de la Escosura (1995), Quiroga and Coll (1998), Sánchez-Albornoz (1987), Simpson (1997) and Tortella (1994).

3 This problem also has economic, political, educational and cultural causes that it is not possible to analyse here.

4 This is an important and complex question that needs a more comprehensive explanation. Some important studies on the topic are: Solow (1957), Kuznets (1962), Gerschenkron (1965), Schmookler (1966), Landes (1969), Rosenberg (1982), Mowery and Rosenberg (1988), Mokyr (1990) and Freeman (1990).

5 For a good explanation of this concept, see Enos (1991).

6 There has been an interesting debate about the validity of patents as indicators of the technology actually generated in countries. See, for example, Kuznets (1962), MacLeod (1988), O'Brien, Griffiths and Hunt (1995), and Sullivan (1990).

7 Figures obtained from Lavín (1997).

8 We have compiled annual series of patents for roughly 30 countries from different sources, including the *Journal of the Patent Office Society* (1964), OMPI (*Organization Mondiale de la Proprété Intellectuelle*, 1983), and *Boletín Oficial de la Propiedad Industrial* (BOPI). The complete compilation has been published in Ortiz- Villajos (1999), Appendix 1.

9 Figure 8.2 shows the data of each country as the percentage of the average value. In absolute terms, all countries increased the number of patent applications throughout the period.

10 As Mar Cebrián has pointed out in a recent seminar in Madrid, this entrepreneurial attitude may also be attributed to the traditional excessive protection of Spanish markets and industries.

11 As, for example, Martín (1999) has pointed out.

12 The *Consejo Superior de Investigaciones Científicas* (High Council for Scientific Research) was created in 1939.

13 A summary and interpretation of this information can be obtained in Ortiz-Villajos (2001).

14 The Council for Extension of Studies was created to promote science activities and research.

15 The National Foundation for Scientific Research.

16 The patents registered in Spain from 1882 to 1935 (Ortiz-Villajos 1999, chs 7 and 8) also show a greater technological dynamism during the 1880s. The rate of high-tech patents was superior in that decade, and the percentage of agricultural patents dropped dramatically after 1891, when high tariffs were placed on grain.

17 Explained in Ortiz-Villajos (2001).

18 Other patent studies containing disaggregated information are: Schmookler (1966), MacLeod (1988), Sokoloff and Khan (1990), Sáiz González (1999).

19 The importance of the countries according to the number of patents was very similar to their importance as capital investors in Spain.

20 In 1907, the participation of the four sectors in the Spanish GDP was: agriculture, 38.2 per cent; manufacturing, 22.5 per cent; construction, 5.8 per cent; and services, 34.0 per cent (Prados de la Escosura 1993, Tables C1, C2 and C4).

21 This indicator must be carefully analysed, as Spain was not a global economy in which all countries were interested. The most respected patent source for comparing the technological levels of countries are the records of the US Patent Office, which is the source used by Lingärde and Saarinen (2002).

22 To obtain this information detailed by countries, see Ortiz-Villajos (2004), Chart 6.

23 This index is analogous to the Revealed Comparative Advantage index of Balassa (1965), used by Crafts (1989). John Cantwell may be the pioneer in the use of patents to calculate RTA. See, for example, Cantwell (1990). Buesa (1992) used this indicator first in Spain.

24 As seen in Table 8.4, this proportion was higher in previous years. The decline of the importance of agricultural patents was likely related to the increase of protectionism from 1891 onwards.

25 This subject is explained in Simpson (1997).

26 This indicator can pose problems, because some patents applied for by individuals could in fact be entrepreneurial, as an individual could apply for a patent on behalf of a company. What it is more certain is that a patent applied for by a company was an entrepreneurial one.

27 According to Carreras and Tafunell (1996), low investment in technology inhibited the emergence of large industrial companies in Spain.

28 Although this argument can be seen as non-scientific, the contemporary problems of Spain cannot be understood without taking its past into account. From 711 to 1492, the main objective of the country as a whole was the reconquest of the Iberian Peninsula, followed by the conquest of America, and the European wars of religion. The monarchy was engaged in foreign campaigns, rather than investing within in order to improve domestic economic and industrial competitiveness (Marcos Martín 2000, para. 3.4).

29 Núñez (1992) has made a complete study of the causes of the low literacy level of Spain and its influence on her economic backwardness.

30 The debate about the scarcity of entrepreneurialism has become heated. This debate was introduced by Gabriel Tortella (1994, ch. 8), who emphasizes the importance of long-term historical factors in creating a particular entrepreneurial mentality in Spain.

References

Aracil, J. and J.L. Peinado (1976), 'Clasificación funcional de los gastos del estado (1850–1965)', in *Datos Básicos para la Historia Financiera de España* (varies authors) (Madrid: Instituto de Estudios Fiscales).

Balassa, B. (1965), 'Trade Liberalization and "Revealed" Comparative Advantage', *Manchester School*, 33, pp. 99–123.

BOPI (*Boletín Oficial de la Propiedad Industrial*: Industrial Property Official Bulletin), 1886–1936 (Madrid: Spanish Patent Office).

Buesa, M. (1992), 'Patentes e innovación tecnológica en la industria española (1967–1986)', in J.L. García Delgado and J.M. Serrano Sanz (eds), *Economía Española, Cultura y Sociedad. Homenaje a Juan Velarde Fuertes* (Madrid: Eudema), pp. 819–55.

Cantwell, J. (1990), 'Historical trends in international patterns of technological innovation', in James Foreman-Peck (ed.), *New Perspectives on the late Victorian Economy: Essays in Quantitative Economic History, 1860–1914* (Cambridge: Cambridge University Press).

Carreras, A. (1992), 'La producción industrial en el muy largo plazo: una comparación entre España e Italia de 1861 a 1980', in L. Prados and V. Zamagni (eds), *El desarrollo económico en la Europa del Sur: España e Italia en perspectiva histórica* (Madrid: Alianza), pp.173–210.

Carreras, A. and X. Tafunell (1996), 'La gran empresa en la España contemporánea: entre el Mercado y el Estado', in F. Comín and P. Martín Aceña (eds) (1996), pp. 73–90.

Cebrián, M. and S. López (2002), 'Economic growth, technology transfer and convergence in Spain, 1960– 1973', in *XIII International Economic History Congress, Session 32* (Buenos Aires: International Economic History Association).

Comín, F. and P. Martín Aceña (eds) (1996), *La empresa en la Historia de España* (Madrid: Civitas).

Crafts, N.F.R. (1989), 'Revealed Comparative Advantage in Manufacturing, 1899–1950', *The Journal of European Economic History*, 18 (1), (Spring), pp. 127–37.

Cubel, A. and J. Palafox (1997), 'El *stock* de capital de la economía española, 1900–1958', *Revista de Historia Industrial*, 17, pp. 113–46.

Enos, J.L. (1991), *The Creation of Technological Capability in Developing Countries* (London: Pinter).

Fagerberg, J. (1987), 'A technology gap approach to why growth rates differ', in C. Freeman (ed.), (1990), pp. 55–67.

Freeman, C. (ed.) (1990), *The Economics of Innovation*, International Library of Critical Writings in Economics, Vol. 2 (Aldershot: Edward Elgar).

Fraile Balbín, P. (1999), 'Industrial Policy under Authoritarian Politics: the Spanish Case', in J. Foreman-Peck and G. Federico (eds), *European Industrial Policy* (Oxford: Oxford University Press), pp. 233–67.

Gerschenkron, A. (1965), *Economic Backwardness in Historical Perspective. A Book of Essays* (New York: Praeger).

Griliches, Z. (1990), 'Patent Statistics as Economic Indicators: A Survey', *Journal of Economic Literature*, XXVIII (December), pp. 1,661–707.

Journal of the Patent Office Society (1964), 'Historical Patent Statistics, 1791–1961', *Journal of the Patent Office Society*, XLVI (February), No. 2.

Kuznets, S. (1962), 'Inventive Activity: Problems of definition and measurement', in Richard R. Nelson (ed.), *The Rate and Direction of Inventive Activity* (Princeton, NJ: Princeton University Press).

Landes, D.S. (1969), *The Unbound Prometheus* (Cambridge: Cambridge University Press).

Lavín, R.R. (1997), '!Que inventen ellos!', *Actualidad Económica*, 27 January, p. 34.
Lingärde, S. and J. Saarinen (2002), 'Technological Specialization in Sweden and Finland 1963–97: Contrasting Developments' (in this volume).
MacLeod, C. (1988), *Inventing the Industrial Revolution. The English Patent System, 1660–1800* (Cambridge: Cambridge University Press).
Marcos Martín, A. (2000), *España en los siglos XVI, XVII y XVIII* (Barcelona: Crítica).
Martín, C. (1999), 'La posición tecnologica de la economía española en Europa. Una evaluación global', *Papeles de Economía Española*, 81, pp. 2–19.
Mokyr, J. (1990), *The Lever of Riches. Technological Creativity and Economic Progress* (Oxford: Oxford University Press).
Mowery, D.C. and N. Rosenberg (1989), *Technology and the Pursuit of Economic Growth* (Cambridge: Cambridge University Press).
Mulet, J. (1998), 'El Sistema Español de Innovación', in A. Ollero, A. Luque and G. Millán (eds), *Ciencia y tecnología en España: bases para una política* (Madrid: FAES).
Núñez, C.E. (1992), *La fuente de la riqueza. Educación y desarrollo económico en la España contemporánea* (Madrid: Alianza).
O'Brien, P.K., T. Griffiths and P. Hunt (1995), 'There is Nothing Outside the Text, and There is No Safety in Numbers: A Reply to Sullivan', *The Journal of Economic History*, 55, pp. 671–2.
OEPM (*Oficina Española de Patentes y Marcas*: Spanish Patent Office) (1882), *Libros de Registro de Patentes (Patent Record Books)*, Madrid.
OMPI (*Organization Mondiale de la Propriéte Intellectuelle*) (1983), *100 Years of Industrial Property Statistics* (Geneva: OMPI).
Ortiz-Villajos, J.M. (1996), 'Patents, Technological Progress and Economic Development, 1791–1993. International Comparisons', paper presented for the European Historical Economics Society Summer School: *Technology and Long-run Growth in Europe, 1500–1990*, Montecatini Terme, Italy, 17–23 June.
Ortiz-Villajos, J.M. (1999), *Tecnología y desarrollo económico en la historia contemporánea. Estudio de las patentes registradas en España entre 1882 y 1935* (Madrid: Oficina Española de Patentes y Marcas).
Ortiz-Villajos, J.M. (2001), 'Instituciones de I+D y progreso tecnológico en la historia de España', in *VII Congreso: Ponencias y Comunicaciones* (Zaragoza: Asociación Española de Historia Económica).
Ortiz-Villajos, J.M. (2004), 'Spanish patenting and technological dependency, pre-1936', *History of Technology*, pp. 24.
Prados de la Escosura, L., T. Daban and J. Sanz (1993), *De Te Fabula Narratur? Growth, Structural Change, and Convergence in Europe, 19th-20th Centuries* (Madrid: Dirección General de Planificación (Ministerio de Economía y Hacienda), Documento de trabajo no. 93009).
Prados de la Escosura, L. (1993), *Spain's Gross Domestic Product, 1850–1990: A New Series*, Ministerio de Economía y Hacienda, Working Paper 93002.
Prados de la Escosura, L. (1995), *Spain's Gross Domestic Product, 1850–1993: Quantitative Conjectures*, Working Paper 95–06 (Madrid: Carlos III University).
Quiroga, G. and S. Coll (1998), 'Another way to look at inequality: income distribution in the mirror of height differences: the case of Spain: 1895–1950', in L. Borodkin and P. Lindert (eds.), *Trends in income inequality during industrialization*, B12 Proceedings 12th International Economic History Congress (Madrid: Fundación Fomento de la Historia Económica), pp. 115–27.
Rosenberg, N. (1982), *Inside the Black Box. Technology and Economics* (Cambridge: Cambridge University Press).

Sáiz González, J.P. (1999), *Invención, Patentes e Innovación en España (1759– 1878)* (Madrid: Oficina Española de Patentes y Marcas).

Sánchez Ron, J.M. (1999), *Cincel, martillo y piedra. Historia de la ciencia en España (siglos XIX y XX)* (Madrid: Taurus).

Sánchez-Albornoz, N. (ed.), (1987), *The Economic Modernization of Spain, 1830–1930* (New York: New York University Press).

Schmookler, J. (1966), *Invention and Economic Growth* (Cambridge, MA: Harvard University Press).

Simpson, J. (1997), *La agricultura española (1765– 1965): la gran siesta* (Madrid: Alianza).

Sokoloff, K.L. and B.Z. Khan (1990), 'The Democratization of Invention During Early Industrialization: Evidence from United States, 1790–1846', *The Journal of Economic History*, 50, pp. 363–78.

Solow, R.M. (1957), 'Technical Change and the Aggregate Production Function', *Review of Economics and Statistics*, XXXIX (August), pp. 312–20.

Sullivan, R.J. (1990), 'The Revolution of Ideas: Widespread Patenting and Invention during the English Industrial Revolution', *The Journal of Economic History*, 50 (June), pp. 349–62.

Tortella, G. (1994), *El desarrollo de la España contemporánea. Historia económica de los siglos XIX y XX* (Madrid: Alianza).

9
Technological Specialization in Sweden and Finland 1963–97: Contrasting Developments[1]

Svante Lingärde and Jani Saarinen

Introduction

Technological differences between countries have been a widely- discussed issue in economic history in recent decades. Theories about early and late adopters of technology and catch-up processes (Denison 1967; Abramovitz 1986), national innovation systems (Freeman 1987; Nelson 1993), and competitive advantages of countries (Dosi, Pavitt and Soete 1990), have been advanced and debated. In new growth theories, the expenditures in R&D have been used as a proxy, along with capital and knowledge, to measure 'inputs' into technological development (OECD 1994a).

The number of patents granted, whether domestic or international, is regarded as a measure of the technology 'output' of countries. It has been stated that the patent statistics of the USA are the best source for assessing global patenting activity (Ray 1988). More specialized research into these patents has helped us to understand both the levels of technology in an international context of different countries, and the specialization of countries in different technological fields. Despite some misleading qualities, patenting data has proven to be an important and informative variable, at least in longer time- series studies. It has also been demonstrated that small and medium-sized countries have a tendency to protect their innovations using patents taken out in foreign countries (Archibugi and Möller 1993). Hence, when two small countries are compared, data on patenting in a third country contributes important information.

The objective of this study is to analyse empirically Swedish and Finnish patenting in the United States Patent and Trademark Office (USPTO), and to advance possible explanations for their specialization in different industrial branches during different periods of time, despite similar resource endowments (cf. Lingärde and Tylecote 1999). In order to achieve this objective, and place it in a context of long-run structural change, we shall use relative 'revealed advantage' measures based on USPTO data along with OECD data on R&D, production and exports. We assume that, during different periods of time, the number of patents granted in a third country (the USA in this case) reflects the technological specialization patterns, as well as the export profile, of the studied country at that particular point in time.

In small countries such as Sweden and Finland, financial, political and other institutions support the growth of certain sectors in order to generate positive effects on other sectors. Although a study focusing on quantities and quantity composition does not acknowledge these linkage effects, we shall try to determine whether the branch profile of patenting has changed internally, or whether institutional (outer) factors have contributed to the change.

After clarifying the branch profile of patenting, we shall turn our focus to how differences in US patenting between Sweden and Finland are related to other differences and shifts in the industrial structures of the two countries and acquire a more nuanced picture of the technological levels and specializations in Sweden and Finland during the studied period.

Earlier studies

To our knowledge, no previous investigation has had the explicit aim of comparing Swedish and Finnish patenting in an economic-historical context. Data on domestic patenting has been used at an aggregate level in historical studies of the inter-play between innovation and growth in the Swedish economy. For example, Krantz (1982) tested a Schumpeterian hypothesis (innovation, as indicated by patenting, causes growth) against a Schmooklerian one (growth causes patenting). For the period from 1835 to the early 1970s, he discerned an alternating pattern in which business cycles in patenting have sometimes been leading, sometimes lagging, in relation to cycles in industrial production at fixed prices. The Schmooklerian pattern, with patents lagging, was dominant from the early 1960s to the early 1970s. Krantz also found that the share of foreign patents among total patents in Sweden rose during periods of a rising ratio of exports to GDP (that is, the post-war period until the 1970s). He hypothesized that Swedish external patenting might be positively correlated with these indicators of international openness, although this hypothesis was not tested.

Papahristodoulou (1986) tested the relationship between, on the one hand, Swedish patents issued and, on the other hand, industrial investments as well as value added. He found some support for the Schumpeterian view that innovation, as indicated by patents, preceded (and perhaps caused) growth. In contrast to Krantz, Papahristodoulou tested for the entire 1896–1982 period without periodization. He also compared the aggregate R&D figures with patents (applications or grants, with different leads and lags), for the 1963–82 period. This investigation of a relatively short time-series (due to the lack of a long R&D series) generated somewhat ambiguous results as regards the lag structure, as well as the strength of the correlation. Thus,

Papahristodoulou concluded that the aggregate patent series must be interpreted with caution when used as an indicator of inventive or innovative activity.

US patents, disaggregated by technological sector, have been used in studies on the development over time of the RTA of various countries, including Sweden and Finland. Originally this index was used in a moderately different form, in trade theory (Balassa 1965). In patent-related studies, the index became widely used after the work done by Soete (1981) and Patel and Pavitt (1991). This specialization index uses the number of patents (p_{ij}) of country i in technology field j and is defined as a country's share of all US patenting in a technological field, relative to its share in all US patenting in all fields. A value above unity indicates a comparative advantage of country i in the technological field j. In other words, the RTA index can be calculated by using the following formula:

$$RTA_{ij} = \frac{p_{ij}/\sum_i p_{ij}}{\sum_j p_{ij}/\sum_i \sum_j p_{ij}}$$

Vertova (1999) analysed RTA, as indicated by patenting in the USA, for eight industrial countries, including Sweden, and for 56 technological sectors for four periods: 1890–1914, 1915–39, 1940–64, and 1965–90. She found that the Swedish RTA in the two post-war periods was correlated positively and significantly (across branches) with the US RTA but not with the RTA profiles of the UK, Germany, France, Switzerland, Italy or Japan. Hence, the branch structure of Swedish US patenting resembled the structure of US domestic patenting rather than the US patenting of the larger European countries. Vertova also found evidence for the tendency, predicted by some 'national systems of innovation' theories, that, in the long run, countries economically converge by specializing in increasingly dissimilar technologies. This was particularly true in cases where two countries or groups of countries were technologically dissimilar from the outset.

Ray (1988) analysed Finnish RTA, again as indicated by patenting in the USA, for 23 technological sectors (based on the US patent classification) and found some signs of a positive correlation between each sector's RTA in the 1981–6 period and the development of its RTA between 1969 and 1974 and 1981 and 1986. Hence, those sectors with a growing RTA also tended to reach a high absolute level of RTA towards the end of the period studied. Several of those progressive sectors started at a low level: that is, they were initially on the weak side of the country's technological specialization.

An important development in Finnish innovation research in recent years has been the publication of a number of studies connected with the compilation and use of the so-called Sfinno database. Close to 3,500 innovations, commercialized from 1945 onwards, have been identified from various sources such as journal articles, companies' annual reports, and expert judgements. The innovations have been classified by product classes and by the industrial sectors of the innovating firms. The connection between innovativeness and economic performance has been explored by Saarinen (2000), who found a positive relationship between innovativeness and growth at the branch level in the 1980s and 1990s.

Both Sweden and Finland, along with other European OECD countries, have been included in the Community Innovation Survey (CIS), which has recorded innovation data for a large sample of firms.[2] It is, however, difficult or even impossible to trace this data very far back in time. In order to study longer periods, we must rely on traditional data, such as patents and R&D expenditures at branch level.

According to a study by Schiffel and Kitti (1978), the development of different countries' US patenting (in 1965–74) was accounted for largely by two factors: the development of each country's domestic patenting, and its goods exports to the USA. However, opinions differ as to the relative importance of these factors and to the direction of causality (cf. Pavitt and Soete 1980; Basberg 1983; 1984). Our study, while focusing on the branch level rather than on aggregate figures, will to some extent be comparable to the study done by Schiffel and Kitti, if domestic R&D expenditures are regarded as a proxy for domestic innovation, and total exports (by branch) as a proxy for exports to the USA. We will not address the question of the direction of causality, however.

It is our hope that the present study will shed light on a question inspired by Vertova's study: namely, whether technological divergence, combined with economic convergence, has also occurred between Sweden and Finland, two countries traditionally showing a high degree of technological and structural similarity in a global perspective.

Methodological issues

Patent numbers can be regarded as an indicator of inventive as well as innovative activity, but they have their pitfalls. The main drawback of patent data as an indicator is that not all inventions and innovations are patented. This is particularly true of process innovations, which makes patent statistics more likely to under-state their role. Some companies – including many small companies – may find patenting too slow or expensive, and attempt to protect their innovations by other measures such as copyrighting or through secrecy (Archibugi 1992; Arundel and Kabla 1998; cf. Basberg 1984, p. 57). It is also true that not all innovations are technically patentable and, conversely, that not all patented inventions become innovations. Some patents are intended for the blocking-out of competitors rather than for immediate commercial exploitation by the assignee. The economic value of patents is not only variable, but 'highly skewed, with those of high value concentrated in a very small percentage of the total' (OECD 1994b, p. 15). This calls for caution in inter- sectoral as well as inter-temporal comparisons. For instance, it has been argued that the average economic value per patent in France and Germany rose systematically during the 1970s (Griliches, Pakes and Hall 1986, p. 24).

Other systematic qualitative differences may occur, as is sometimes revealed through analysis of patent citations (citations of existing patents in new patent applications). From 1975 to 1980, Swedish US patents in laser technology were less cited, on average, than US patents registered by Germany and Japan in the same field (Granberg 1986). (It is possible, however, that this was due to bias among large German and Japanese companies, who opted to cite their own patents rather than those of their competitors.) We do not know of any similar comparison between Swedish and Finnish US patents.

Patent counts are also vulnerable to disturbances such as the slump in granted US patents in 1979, caused by a lack of funds for the printing of patents (OECD 1994b, p. 41). Other examples of institutionally-generated changes in total US patenting include the improved success of some companies in asserting their patents in court in the latter half of the 1980s, and the lengthening of the patent term in 1995 (Hicks *et al.* 2001). Relative measures, such as RTA (which we are using), are probably less sensitive to such disturbances.

Patents are often regarded as an output indicator of the innovative process, whereas R&D investments – which are also studied in this chapter – are regarded as an input indicator (OECD 1994a). This interpretation should be used with caution, however. As they appear in official statistics, R&D investments do not constitute an entirely reliable measure of technological development efforts. As Laestadius (1996; 2000) notes, the definition of R&D in OECD's Frascati Manual tends to favour the collection of data on knowledge creation of a more scientific nature, at the expense of other types of creative activity of a more synthetic or integrative nature. Furthermore, large companies tend to be over-represented (Jacobsson, Oskarsson and Philipson 1996).

In addition, time-series studies do not always warrant the interpretation that R&D expenditures constitute the leading variable and patents the lagging one, or that changes in R&D expenditures affect the number of patents in terms of Granger causality (Hall, Griliches and Hausman 1986). The nominal R&D expenditures of a firm may peak either early or late in the innovative process, depending on the nature of the firm, the branch of industry, and the technology. Similarly, the time-lags between a technical invention and a patent application, and between a successful application and a granted patent, differ over time, between branches, and (especially in the case of domestic patents), between countries. Hence in some cases patents may be granted to a company before the development costs for the new product or process reach their peak, and similar phenomena may be seen at branch level.

Recent international debates on the falling ratio of granted patents to R&D costs, as well as modern models of the innovation process, have highlighted the problems of simple input–output conceptions of R&D and patenting (Griliches, Pakes and Hall 1986; Kleinknecht and Bain 1993; Niininen and Saarinen 2000). In this study we do not attach importance to the leads or lags between R&D and patents, but focus on the extent to which similarities and differences between Sweden and Finland in R&D series, as well as export and production series, are matched in the corresponding patent series. The differing long-term trends of R&D expenditures and patenting in certain branches constitute an exciting starting-point for the study, rather than a peripheral problem.

At the branch level, classification and concordance problems are inevitable (Soete 1987). When patent data are compared with R&D and other economic data, the classic problem for the researcher is that patent statistics are disaggregated by patent (technology) classes, whereas economic statistics are disaggregated by industrial branches or products. In the case of patent statistics the unit of analysis is the patent, while for other statistics (such as R&D statistics) the company or the plant constitutes the account unit. Hence, R&D performed by a company in a certain branch may well result in patents that, by way of the patent classification and the USPOC-SIC (see below) concordance, are associated with other branches.

There are standard concordances between the most important patent classification systems (International Patent Classification or IPC, and the United States Patent Office Classification or USPOC) and the International Standard Industrial Classifications (ISIC and SIC: see the Appendix to this chapter). However, their validity depends on whether we want to subdivide patent numbers by the probable supplier industries or the probable user industries, or both (Griliches 1990). The standard concordance between the USPOC and the SIC, which is used in the statistics we have derived from the US Patent and Trademark Office (see below), may to some extent be seen as a compromise in which product inventions tend to be registered by the probable supplier industries, and process inventions by the probable user industries.

US patenting of Sweden and Finland: an overview

Our source of patent data is the USPTO. We have analysed the number of US patents granted to Sweden and Finland during the 1963–97 period. The period under investigation in our study is the period covered by the Technology Assessment and Forecast (TAF) database provided by the USPTO, in which patents are searchable by industrial (SIC) class. The rationale for using patent data in our study is that, despite the problems mentioned above, patents reflect continuous developments within technology (Engelsman and van Raan 1990). This is especially useful since our study intends to analyse differences in patenting activity over a long period (almost 40 years). Therefore, the time lags that exist in the case of patent indicators do not play a critical role in the long-term developments under study. Our reason for using US patent data is its high degree of consistency over time and between countries of origin (Soete and Wyatt 1983).

We will begin this empirical part of the study by introducing data concerning the patenting activity of our studied countries. When we have used the patent data, we have concentrated on the application year instead of the grant year. This has been done in order to avoid the time-lag, which lasts at least two years from the date of application to the date of the grant, depending on the industrial branch. Focusing on the application date provides better information about the precise point in time when the new knowledge was generated and transferred. In Figure 9.1, the development of US patenting of Sweden and Finland, according to the date of application, is outlined.

As Figure 9.1 illustrates, there are considerable differences in US patenting between Sweden and Finland. Sweden experienced a rapid increase in number of patents at the beginning of the studied period. In 1974, growth stagnated and then went into decline until the early 1990s. Since then, growth has accelerated, and the earlier record, in numbers of patents, was surpassed in 1994. (The short decline at the end of the time- series is explained by the time-lag qualities of our data.)

In the Finnish case, the number of patents began at a low level but has shown a clear increasing trend throughout the period. The most rapid growth was experienced in the 1990s, due mainly to the telecommunications sector. In per capita terms, Finland had almost reached the Swedish level by the late 1980s. During the last decade of the period, the development of the two countries' patenting was quite similar. Both countries experienced rapid

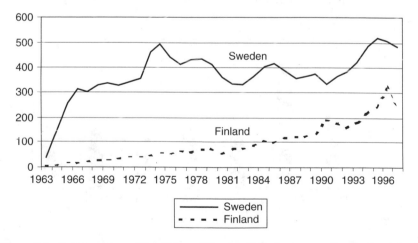

Figure 9.1 Total number of Swedish and Finnish US patents according to the date of application, 1963–97

growth in US patenting, caused mainly by the electronic industries. As we shall see, however, this similarity at the aggregate level was not generally found at branch level.

Differences in the branch profiles of patenting

The variance of RTA across branches can be regarded as a measure of the strength of the country's technological advantages and disadvantages. The variance of RTA values is shown in Figure 9.2.

The graph reveals that the variance among branches has been much greater in Finland than in Sweden during the period under study. It should be noted that the RTA variance is a biased measure that tends to be higher for countries with smaller numbers of patents, although this does not account entirely for the disparity between Sweden and Finland.[3] It should also be noted that RTA variance may be sensitive to the chosen level of aggregation. The disaggregation could have been more sensitive within several categories, including 'chemicals not otherwise specified' (soaps and so on, paints and so on, and miscellaneous), 'electrical equipment not otherwise specified' (household appliances, lighting and wiring, and miscellaneous), and 'transportation equipment' (with eight subcategories). But it turns out that the RTA variances *within* these groups were high for both countries throughout much of the period (despite small number problems), and thus the pattern of a generally higher RTA for Finland remains, regardless of the aggregation level. Therefore, what we can state is that RTA variance across branches has been systematically greater for Finland than for Sweden, and that the trend during the last decade has been one of increased variance in Finland, while the RTA variance of Sweden has been quite stable and has even diminished.

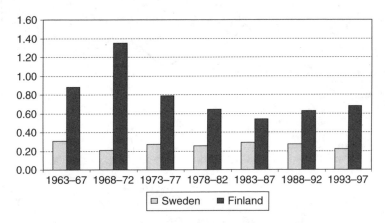

Figure 9.2   Variance in revealed technological advantage in Sweden and Finland, 1963–97

To account for the relatively even distribution of patents in Sweden, one might argue that the different industrial branches were almost equally developed at the beginning of the 1960s, and shared a similar development path during the period studied. It is well-known that the Swedish economy is heavily dependent on exports; thus the many-sidedness of Swedish US patenting may be an indicator of the export orientation in the industry as a whole.

In the case of Finland, a hypothesis of the relationships between patenting, trade, and structural change may be formulated as follows: the high value of the RTA variance may be explained by the important role played by mechanical engineering industries in Finnish exports to the USA until the 1980s. Forest-based industries were even more important in total Finnish exports during these decades, but they did not have any markets in the USA. In the 1980s, electrical and electronics industries caught up with mechanical industries, which explains the decline of the RTA variance. For the latest years, then, it might be seen how the electrical and electronics branches have become the motor of industrial development in Finland, which again has increased RTA variance.

In Table 9.1, the RTA figures for all studied sectors for Sweden and Finland are presented. The period has been divided into two subperiods, from 1968 to 1987, and from 1988 to 1997. The RTA figures in parentheses indicate that the number of patents in that sector during the subperiod was less than 20. The bold numbers indicate whether a country has a technological advantage in a selected technology class. We will comment further on these developments later, when comparing RTA with other variables.

Comparison of patenting: R&D, production and exports

To what extent are country differences and development patterns in RTA matched in R&D expenditures, production, and exports? In order to study these variables, we have used an OECD data set with annual data for a shorter period (1980 to 1996).

If a country's export profile affects the profile of its US patenting, a correlation between RTA and exports should be expected. Similarly, if domestic innovation affects US patenting, it is reasonable to expect RTA to be correlated with R&D expenditures. We have no particular expectations for time-lags in either direction. Since our analysis covers long-term structural change, we will also include an indicator of total branch output to see if changes in the composition of production match the changes in exports, R&D and patenting.

Since RTA is a relative measure, in which each branch of manufacturing in a country is compared with other branches, as well as with other countries, we will use similar relative indicators of R&D, exports and production. As an

Table 9.1 Revealed technological advantage indices for Sweden and Finland

	Food and kindred products [01]		Textiles, apparel & leather [02]		Industrial inorganic chemistry [06]		Industrial organic chemistry [07]		Plastics materials & synthetics [08]	
	Swe.	Fin.	Swe.	Fin.	Swe.	Fin.	Swe.	Fin.	Swe.	Fin.
1968–1987	0.95	(1.38)	0.72	0.64	0.84	1.97	0.43	0.37	0.11	(0.24)
1988–1997	1.42	1.48	0.62	0.89	1.10	1.88	0.70	0.66	0.11	0.60

	Agricultural chemicals [09]		Soaps, paints & misc. chemicals [10]		Drugs and medicines [14]		Petroleum & gas extraction [15]		Rubber & misc. plastic products [16]	
	Swe.	Fin.	Swe.	Fin.	Swe.	Fin.	Swe.	Fin.	Swe.	Fin.
1968–1987	0.91	(0.37)	0.56	(0.36)	1.08	(2.51)	0.29	(0.27)	1.15	0.90
1988–1987	1.20	0.85	0.55	(0.31)	1.03	(0.96)	(0.37)	(0.21)	1.01	0.65

	Stone, glass & concrete products [17]		Primary ferrous products [19]		Prim. & second. non-ferrous metal [20]		Fabricated metal products [21]		Engines & turbines [23]	
	Swe.	Fin.	Swe.	Fin.	Swe.	Fin.	Swe.	Fin.	Swe.	Fin.
1968–1987	1.20	1.26	1.86	(1.43)	1.27	(1.09)	1.31	1.19	1.13	(0.48)
1988–1997	1.31	1.92	1.70	(1.12)	(0.64)	(1.17)	1.37	1.05	1.29	0.73

	Farm & garden machinery [24]		Mining & construction machinery [25]		Metal-working machinery [26]		Office computing machines [27]		Special industry machinery [29]	
	Swe.	Fin.	Swe.	Fin.	Swe.	Fin.	Swe.	Fin.	Swe.	Fin.
1968–1987	1.77	2.19	1.99	2.60	1.43	1.64	0.64	(0.27)	1.40	3.01
1988–1997	1.58	1.72	1.86	2.88	1.59	1.50	0.42	0.27	1.85	3.36

	General industry machinery [30]		Refrigeration & service machinery [31]		Misc. machinery, except electrical [32]		Electrical trans. equipment [35]		Electrical ind. apparatus [36]	
	Swe.	Fin.	Swe.	Fin.	Swe.	Fin.	Swe.	Fin.	Swe.	Fin.
1968–1987	1.48	1.17	1.07	(1.11)	(1.22)	(0.66)	0.77	0.55	1.12	(0.31)
1988–1997	1.68	1.32	(1.15)	(0.94)	(0.64)	(1.29)	0.95	1.59	0.64	0.83

	Household appliances [37]		Radio & TV equipment [42]		Electronic components etc. [43]		Transportation equipment [44]		Prof. & scientif. instruments [55]	
	Swe.	Fin.	Swe.	Fin.	Swe.	Fin.	Swe.	Fin.	Swe.	Fin.
1968–1987	0.90	0.78	0.53	(0.17)	0.62	0.46	1.69	2.12	0.84	0.59
1988–1997	1.29	0.65	1.21	2.31	0.76	0.87	1.28	(0.30)	0.91	0.60

export indicator, we will construct a Revealed Comparative Advantage, RCA, which is calculated according to the following formula:

$$RCA_{ij} = \frac{X_{ij}/\sum_i X_{ij}}{\sum_j X_{ij}/\sum_i \sum_j X_{ij}}$$

where X_{ij} denotes the value of exports of country i in manufacturing branch j.[4]

As an R&D indicator, we construct an analogous index, which we call the Revealed R&D Advantage, RRDA:

$$RRDA_{ij} = \frac{RD_{ij}/\sum_i RD_{ij}}{\sum_j RD_{ij}/\sum_i \sum_j RD_{ij}}$$

where RD_{ij} denotes R&D expenditures in branch j in country i.

Finally, we construct an index of the Revealed Production Advantage, RPA:

$$RPA_{ij} = \frac{Q_{ij}/\sum_i Q_{ij}}{\sum_j Q_{ij}/\sum_i \sum_j Q_{ij}}$$

where Q_{ij} is the output of branch j in country i.

The i sums refer to OECD country groups whose composition differs somewhat according to data availability, so that there is no perfect comparability between the indices. However, for each index, there is comparability over time between our two countries. The j sums refer to the whole manufacturing industry.

In the next phase of our analysis we combine Sweden and Finland in terms of our newly-constructed indicators RCA, RRDA and RPA. The broadly similar industrial structure of Finland and Sweden implies that we could expect a similar pattern for the different indices. However, the results from the previous section suggest some increasing technological differences in the branch profile of patenting. Therefore, we expect to see similar development in the other indicators also, because changes in the technological specialization of countries cannot be caused by technology alone. Market forces, science and technology policy and other related factors play important roles in that process. Figure 9.3 illustrates the correlations between our three indices – exports, R&D, and production – across 22 branches for the years 1980, 1990, 1994 and 1996.

The fall in all three correlation coefficients provides striking evidence of a specialization divergence between Sweden and Finland between 1980 and 1996.[5] Correlation coefficients are consistently positive, as we should expect from two structurally-similar economies. But obviously, this similarity decreased considerably during the studied period.

The major decrease occurred between 1990 and 1994 as regards R&D advantage and production advantage. The two countries' revealed comparative advantage remained fairly strongly correlated in 1994, but this correlation had decreased dramatically by 1996. A study of how each country's

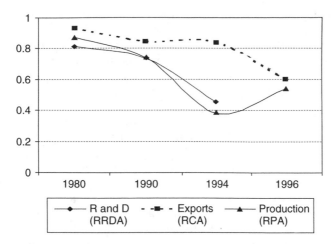

Figure 9.3 Correlations for Sweden and Finland

Source: Based on data from OECD (1999).

revealed exports, R&D and production index is correlated with itself at different points in time (not shown here) reveals that the greatest changes
between 1990 and 1994 took place in Sweden rather than in Finland. The
outcome of the economic crisis of the early 1990s was different in Sweden,
where the comparatively strong and continuous R&D growth appears to
have been relatively independent of the slump in production. Finland, on
the other hand, seems to have developed in tandem with other OECD countries in structural terms, despite a greater slump in production. In exports,
the structural changes within each country were not dramatic. But in the
Finnish case, much of the change was concentrated in the last two years
of the period (1994–6), with new export products finding markets after the
slump in exports to Russia that hit Finnish industry hard. This helps to
explain why the Swedish–Finnish RCA correlation weakened in those years.

Interestingly, divergence between Sweden and Finland is much less marked
in RTA than in the other indices. RTA levels cannot be compared over a two-
year period, due to the small annual number of patents at the sectoral level.
But if we compare the 1968–87 and 1988–97 periods, the Swedish–Finnish
RTA correlation declines only marginally: from 0.73 in the earlier period to
0.62 in the later period. A second observation is that Finland's patenting
profile has apparently changed much more than the profile of Sweden. The
cross-section correlation between the earlier and the later period is 0.80 for
Sweden, but only 0.33 for Finland. When the other variables are taken into
account, however, the picture is completely different. If we compare the
years 1980 and 1994 (as well as subperiods between those years), Finland

shows much greater stability: that is, higher correlation between the periods in R&D, exports, and production indices.

How can we interpret this? In one sense, US patenting differs from the other variables in its behaviour. There may be several reasons for this: the periodization is not comparable, or the analysis of RTA suffers from small number problems (despite our branch grouping and periodization), or the difference is 'real'.

We will remain open to all three possibilities in the following discussion. However, if we compare the variables with each other, we find a reasonably strong correlation between US patents and R&D (Table 9.2). This may indicate that our periodization makes sense, since there appears to be comparability between patenting in 1968–87 and R&D in 1980, and between patenting in 1988–97 and R&D in 1994. However, RTA is generally *not* strongly correlated with the other variables: the export profile as indicated by RCA, and the production profile as indicated by RPA. This means that, as far as correlation across branches is concerned, domestic R&D seems to be more influential than exports in determining the level of patenting in the USA. Curiously, production also shows a stronger correlation with US patenting than exports. This phenomenon must have something to do with competition, not only in exports, but also in imports.

We shall now, at branch level, compare the development of RTA between 1968–87 and 1988–97, and the levels and changes in R&D, exports and production between 1980 and 1996 (1994 in the case of R&D). In order to achieve reasonable comparability between RTA and the other indices as regards branch classification, we shall employ a slightly higher level of aggregation than used previously. We divide manufacturing industry into 21 branches, for which we have numbers for R&D, exports and production; RTA numbers are only available for 16 of them. This leaves us with a small residual class, which we shall leave to one side due to problems with index comparability. In particular, it may include some activities within the wood, pulp and paper industries in the case of RTA, but not in the case of the other indices.

Table 9.2 Correlation between RTA and other indices

RTA 1968–87	RRDA 1980	RCA 1980	RPA 1980
Sweden	*0.610	0.317	0.422
Finland	0.182	−0.212	0.090
RTA 1988–97	**RRDA 1994**	**RCA 1994**	**RPA 1994**
Sweden	*0.549	0.322	*0.614
Finland	*0.878	0.181	0.353

* Significant at 5% level ($N = 16$).

For each of the branches, this gives us four different indices showing whether the branch keeps, gains or loses a comparative advantage in patenting, R&D, exports and production, respectively. We shall say that a branch *gains* a comparative advantage either (a) if the index changes from a value below 1 to a value above 1, or (b) if the ratio between the new and the old value is higher than 2, provided that the new value is not below 0.5. Similarly, we shall say that a branch *loses* a comparative advantage either (a) if the index changes from a value above 1 to a value below 1, or (b) if the ratio between the new and the old value is lower than 0.5, provided that the new value is not above 2. If there is no such change, the branch will be said to be consistently 'strong' (if its value is above 1 and does not change substantially), or consistently 'weak' (if its value is below 1 and does not change substantially). Naturally, it should be kept in mind that the labels 'strong' and 'weak', as well as 'gaining' and 'losing', relate only to the position of that branch in relation to other branches in the same country.

If our four indices are considered together, we discern a pattern in which the branches fall into five groups:

(a) consistently strong in both countries (2 branches);
(b) consistently weak in both countries (1 branch);
(c) converging (4 branches);
(d) diverging (5 branches);
(e) irregular or uncertain developments (6 branches).

In the following sections, we will examine the developments in the major parts of these branches. The last group, irregular or uncertain developments, is not considered, owing to the lack of any common characteristics in the development of these branches.

Strong and weak branches

We find two consistently strong branches: *wood products and furniture* and *paper, paper products, and printing*. At present we have no separate patent data for these branches, so the technological index cannot be calculated. Not surprisingly, both branches have been important in both Sweden and Finland in R&D, as well as exports and total production. However, their relative importance has decreased somewhat, especially in Sweden, over the 1980–96 period.

Considering the group of consistently weak branches, *professional and scientific instruments* show low index figures for both countries in patenting, production and exports. The R&D index fluctuates around one in Finland, and the corresponding Swedish figure converges to the Finnish level in the early 1990s. Despite investments in knowledge-based society during the first half of the 1990s, the results are not evident in the development of this particular branch.

Converging branches

Among converging branches, we can identify three cases where Finland converges upwards, and one case where Finland converges downwards. There are no clear examples of Sweden converging on Finland.

In *textiles, apparel and leather*, the indices of US patenting, exports, and production are low, and falling, in both countries throughout most of the period. This is mainly the result of the emergence of competition from low-wage countries, which affected the Swedish and Finnish textile industries severely. This branch can be seen as an example of Finland converging downwards. There are no clear counterparts to this development among the other branches.

In *metal products* (other than machines, electrical equipment, transport equipment and instruments), both countries sit slightly above the unit line in RTA in 1968–87 and 1988–97. Considering the R&D index, developments are sporadic. In Sweden, large investments in R&D began to decline in the mid-1980s, as new industrial branches (particularly ICT) emerged. In Finland, metal products maintained their strong position throughout that decade. It may make sense to refer to the Finnish R&D pattern lagging five to ten years behind the Swedish figure, since some convergence takes place simultaneously in production and exports, with Finland reaching the unit line during the latter half of the 1980s.

In the important branch of *non-electrical, non-office machine industry*, Sweden enjoys clear comparative advantages internationally throughout the period. The Finnish production and export indices show a tendency towards convergence upwards, towards the Swedish level. In R&D, on the other hand, both countries find themselves at an exceedingly high level during most of the 1980s, declining slightly in the 1990s. In an international comparison, the non-electrical, non-office machine industry is more technology-intensive in Sweden and Finland than the behaviour patterns of these countries' manufacturing sectors would lead us to expect. For instance, both Swedish and Finnish patenting and R&D figures in this branch are generally higher than their export and production figures.

Fourth, the branch producing *radio and television receiving equipment, electronic components, and communications equipment* provides another example of Finland converging upwards. In both Sweden and Finland, the potential future markets of this industrial branch were recognized early by companies. Due largely to the big players in both countries (namely Ericsson in Sweden and Nokia in Finland), the development of this branch was rapid. In Finland, Nokia's share of industrial R&D expenditures was some 50 per cent in the late 1990s. Indeed, the story of Finnish convergence in this area is essentially the story of Nokia's convergence.

Diverging branches

As might be expected from the falling trend of correlation coefficients (see Figure 9.3), there are also clear cases of Swedish-Finnish divergence at branch level. We have placed five branches in this category.

In *chemicals except drugs and medicines*, the tendency towards divergence is visible in the export and R&D indices, and to some degree in production. The branch is quite broad, and it does not have any large players in the countries studied. Both countries are marked by comparative disadvantage during the whole period, although Finland's comparative position improves while Sweden's deteriorates. RTA, which increases slightly in both countries, is the only index showing no divergence in the branch as a whole. When RTA is disaggregated more finely, it can be seen that both countries improve their comparative position in industrial organic chemistry and agricultural chemicals. Sweden also improves its comparative position in industrial inorganic chemistry, while Finland shows an improvement in plastic materials and synthetic resins.

In *drugs and medicines*, RTA comparisons are difficult to make. This is due in part to low absolute patent numbers, as well as legal changes during the period concerning the scope of domestic patentability, which may have affected Swedish and Finnish US patenting indirectly. The other indices show divergence between Sweden and Finland. One explanation for the diverging pattern is the innovative and economic success of large Swedish pharmaceutical firms during the last decades.

In *rubber and plastic products*, we can, with some hesitation, speak of a divergence in RTA: it is above the unit line for Sweden and below the unit line for Finland; it falls in both countries but more steeply in Finland. The export index diverges by rising in Sweden and falling in Finland. Both countries show low and slightly falling production indices.

Another type of divergence occurs in *non-metallic mineral products* (stone, clay, glass and concrete products). Here the Finnish RTA increases strongly, particularly during the late 1970s due to a construction boom. In both countries this branch can be seen as a typical example of an industry that invests little in product development and mainly produces for the domestic market. This pattern is also displayed by the data on R&D and exports.

The developments in *transport equipment* are easier to summarize. There is a marked divergence in RTA, where Finland moves from a strong advantage to a strong disadvantage while Sweden keeps a modest advantage. We have preferred not to break down the RTA figure to a more disaggregate level as this would result in problems with low absolute patent numbers.

The other indices, which are more finely disaggregated, also show divergence between the two countries in shipbuilding and aircraft, and non-convergence in motor vehicles. The production of motor vehicles flourished in Sweden due to two large automobile manufacturers, Saab and Volvo. The former has also been actively involved in the aircraft industry. In Finland, motor vehicles as well as aircraft industries are practically non-existent. The shipbuilding industry, on the other hand, has been declining in Sweden but remains prominent in Finnish production and exports. Shipbuilding, as well as car manufacturing and aircraft, are extremely R&D-intensive branches.

Conclusions

We began this chapter with an overview of Swedish and Finnish US patenting during the 1963–97 period, followed by more specialized estimations and calculations. We introduced the revealed technological advantage index, made some correlation estimations, and compared the RTA variance between the two countries. These investigations produced new data on the behaviour and development of Swedish and Finnish industries in terms of their US patenting. Finally, we compared the RTA index with other indices of the technological structure: revealed R&D advantage, revealed comparative advantage (based on exports) and revealed production advantage. Our tentative conclusions are as follows.

The patenting activities of Swedish and Finnish industries have shown considerable differences during the last four decades. Over the whole period, Sweden was granted more patents in the USA than Finland. However, from 1966 to 1997, the number of Swedish US patents has increased slowly and sporadically. In fact, the number of US patents was nearly identical in 1966 and 1990. In Finland, US patenting experienced continuous growth from 1963 until the early 1990s. In per capita terms, Finland caught up with Sweden by the late 1980s. Since 1990, the growth has been rapid in both countries, while the ratio between their US patenting has not changed remarkably.

In Finland, the RTA variance across branches has been much greater than in Sweden. This means that in Sweden most industrial branches have been relatively similar in their patenting behaviour. In Finland, the situation has been very different. According to our results, there have been considerable differences between the industrial branches. The RTA variance decreased until the mid-1980s; since then, it has begun to increase again.

Swedish–Finnish similarities in RTA are mainly found in industries in which RTA has not changed much over time in either country. On balance, there is a minor tendency towards divergence over time in the RTA profiles of the two countries.

In the other indices showing revealed comparative advantages in R&D, exports, and production, Swedish–Finnish correlation decreased between 1980 and 1994, indicating a divergence in their profiles. When investigating the comprehensive picture given by the four indices (including RTA), branch by branch, we found four tentative cases of convergence and five cases of divergence among the 18 branches (that is, cases in which reasonable agreement was found between the different indices). Statistically, there was agreement between RTA and the other indices to such an extent that RTA seems to be especially relevant as a technology indicator in analyses of long periods, despite problems with small absolute numbers of US patents in some branches, particularly in Finland's case.

Overall, there are some signs of divergence between the two countries in the branch patterns of US patenting between 1968–87 and 1988–97, and in the branch patterns of R&D expenditures, exports, and production between 1980 and 1994–96. Although the picture is not unequivocal, we conclude that Finland's catching up with Sweden in aggregate terms, economically and technologically, has taken place at the same time as technological and structural differences between the two countries have increased. This may be interpreted as a sign of the growing importance of specialization in an increasingly globalized economy. Specialization is likely to occur not only at country level but also regionally and locally. New cluster and network formations have become crucial in our two sparsely-populated countries, parallel to the continuing dominance of large plants and corporations. Future research at a disaggregate level, along with updated data and the addition of more variables such as domestic patenting, may shed more light on this.

Notes

1 We are grateful to Jonas Ljungberg, Anders Granberg, Keith Pavitt, Martin Andersson, Joakim Appelquist, Ken Sokoloff, seminar participants at the VTT (Espoo), the Department of Economic History (Lund), and at the International Economic History Congress in Buenos Aires, participants of the Klinta Workshop, as well as an anonymous referee, for their valuable comments on earlier versions of this chapter. Thanks are also due to Riksbankens Jubileumsfond and Ruben Rausings Fond for financial support. All responsibility remains with the authors.

2 For an interesting investigation based on this data, see Lööf *et al.* (2001).

3 Between 1976 and 1993, our general picture of dispersion or specialization for Swedish and Finnish US patenting agrees roughly with a χ^2 measure employed by Pianta and Meliciani (1996). This measure has a similar bias, although, according to the authors, adjusting for the bias makes little difference on the whole. In addition, despite the fact that the number of Finnish patents in 1993–7 (1,199 patents) exceeds the Swedish figure for 1963–7 (1,048 patents), Finnish RTA variance never falls to levels anywhere near the level consistently shown by Sweden.

4 Here and in the following formulae, the values are in current US dollars. Deflating procedures are not needed due to the use of relative measures. Exported commodities are classified by industrial sector according to a standard conversion matrix used by the OECD. This means that the export classification may not be exactly comparable to the classification of production and R&D expenditures, but we assume this problem to be of minor importance. For similar reasons, we have also chosen to ignore the fact that exports and production are converted to US dollars using market exchange rates, while conversion of R&D figures is done with purchasing power parities.

5 Twenty branches for RCA and RPA in 1996 (aircraft and various other transport equipment branches, other than shipbuilding and motor vehicles) have been excluded due to lack of Finnish data. RRDA has not been calculated for this year due to lack of R&D data. The data for 22 branches give full coverage of manufacturing industry.

References

Abramovitz, M. (1986), 'Catching Up, Forging Ahead and Falling Behind', *Journal of Economic History*, 46(1), pp. 217–43.

Archibugi, D. (1992), 'Patenting as an indicator of technological innovation: a review', *Science and Public Policy*, 19(6), pp. 357–68.

Archibugi, D. and K. Möller (1993), 'Monitoring the Technological Performance of a Small Economy Using Patent Data: The Case of Denmark', *Technology Analysis and Strategic Management*, 5(2), pp. 99–113.

Arundel, A. and I. Kabla (1998), 'What percentage of innovations are patented? Empirical estimates for European firms', *Research Policy*, 27, pp. 127–41.

Balassa, B. (1965), 'Trade Liberalization and Revealed Comparative Advantage', *The Manchester School of Economic and Social Studies*, 33, pp. 99–123.

Basberg, B. (1983), 'Foreign Patenting in the USA as a Technology Indicator: The Case of Norway', *Research Policy*, 12, pp. 227–37.

Basberg, B. (1984), *Patenter og teknologisk endring i Norge 1840–1980* (Bergen: Norges Handelshoyskole).

Denison, E.D. (1967), *Why Growth Rates Differ* (Washington, DC: The Brookings Institution).

Dosi, G., Pavitt, K. and L. Soete (1990), *The Economics of Technical Change and International Trade* (Hemel Hempstead: Harvester Wheatsheaf).

Engelsman, E.C. and A.F.J. van Raan (1990), *The Netherlands in Modern Technology: A Patent-Based Assessment* (Leiden: Centre for Science and Technology Studies (CWTS), Research Report to the Ministry of Economic Affairs).

Freeman, C. (1987), *Technology Policy and Economic Performance: Lesson from Japan* (London: Pinter).

Granberg, A. (1986), 'A Bibliometric Survey of Laser Research in Sweden, West Germany, and Japan', *Research Policy Studies*, 172.

Griliches, Z. (1990), 'Patent Statistics as Economic Indicators: A Survey', *Journal of Economic Literature*, 28(4), pp. 1661–707.

Griliches, Z., A. Pakes and B. Hall (1986), 'The Value of Patents as Indicators of Inventive Activity', *NBER Working Paper*, No. 2083.

Hall, B., Z. Griliches and J. Hausman (1986), 'Patents and R and D: Is there a Lag?', *International Economic Review*, 27(2), pp. 265–83.

Hicks, D., T. Breitzman, D. Olivastro and K. Hamilton (2001), 'The changing composition of innovative activity in the U.S. – a portrait based on patent analysis', *Research Policy*, 30(4), pp. 681–703.

Jacobsson, S., C. Oskarsson and J. Philipson (1996), 'Indicators of technological activities – comparing educational, patent and R&D statistics in the case of Sweden', *Research Policy*, 25(4), pp. 573–85.

Kleinknecht, A. and D. Bain (eds) (1993), *New Concepts in Innovation Output and Measurement* (London: Macmillan).

Krantz, O. (1982), Teknologisk förändring och ekonomisk utveckling i Sverige under 1800- och 1900-talen. Iakttagelser från patentstatistiken, *Meddelande från Ekonomiskhistoriska institutionen Lunds universitet*, No. 26.

Laestadius, S. (1996), *Är Sverige lågteknologiskt? – reflektioner kring kunskapsbildning och kompetens inom industriell verksamhet* (Stockholm: KTH, Institutionen för industriell ekonomi och organisation).

Laestadius, S. (2000), 'Biotechnology and the potential for a radical shift of technology in the forest industry', *Technology Analysis and Strategic Management*, 12(2), pp. 193–212.

Lingärde, S. and A. Tylecote (1999), 'Resource-Rich Countries' Success and Failure in Technological Ascent, 1870– 1970: The Nordic Countries versus Argentina, Uruguay and Brazil', *Journal of European Economic History*, 28(1), pp. 77–112.

Lööf, H., A. Heshmati, R. Asplund and S.-O. Nåås (2001), 'Innovation and Performance in Manufacturing Industries: A Comparison of the Nordic Countries', *SSE/EFI Working Paper Series In Economics and Finance*, No. 457.

Nelson, R. (ed.) (1993), *National Systems of Innovation: A Comparative Study* (Oxford: Oxford University Press).

Niininen, P. and J. Saarinen (2000), 'Innovations and the Success of Firms', *VTT Group for Technology Studies Working Paper*, No. 53/00.

OECD (1994a), *Main Definitions and Conventions for the Measurement of Research and Experimental Development (R&D) – A Summary of the Frascati Manual 1993* (Paris).

OECD (1994b), *Using Patent Data as Science and Technology Indicators. Patent Manual 1994* (Paris).

OECD (1999), *Main Industrial Indicators* (Paris).

Papahristodoulou, C. (1986), *'Inventions, Innovations and Economic Growth in Sweden: An Appraisal of the Schumpeterian Theory'* PhD dissertation, Uppsala University, Department of Economics.

Patel, P. and K. Pavitt (1991), 'Europe's Technological Performance', in C. Freeman, M. Sharp and W. Walker, *Technology and the Future of Europe: Global Competition and the Environment in the 1990's* (London: Frances Pinter).

Pavitt, K. and L. Soete (1980), 'Innovative Activities and Export Shares: Some Comparisons between Industries and Countries', in K. Pavitt, (ed.) *Technical Innovation and British Economic Performance* (London: Macmillan).

Pianta, M. and V. Meliciani (1996), 'Technological Specialization and Economic Performance in OECD Countries', in *Technology Analysis and Strategic Management*, 8(2), pp. 157–75.

Ray, G.F. (1988), 'Finnish Patenting Activity', *ETLA Discussion Papers*, No. 263 (Helsinki).

Saarinen, J. (2000), 'Innovation Activity in Finnish Industries – A New Pattern', *Lund Papers in Economic History*, No. 69.

Schiffel, D. and C. Kitti (1978), 'Rates of Invention. International Patent Comparisons', *Research Policy*, 7(4), pp. 324–40.

Soete, L. (1981), 'A General Test of Technological Gap Trade Theory', *Weltwirtschaftliches Archiv*, 117 (), pp.

Soete, L. (1987), 'The impact of technological innovation on international trade patterns: The evidence reconsidered', *Research Policy*, 16, pp. 101–30.

Soete, L. and S. Wyatt (1983), 'The Use of Foreign Patenting as an Internationally Comparable Science and Technology Output Indicator', *Scientometrics*, 5, pp. 31–54.

Vertova, G. (1999), 'Stability in National Patterns of Technological Specialisation: Some Historical Evidence from Patent Data', *Economics of Innovations and New Technology*, 8(4), pp. 331–54.

Appendix: Standard Industrial Classification (SIC; TAF sequence numbers) with Swedish–Finnish correlation, 1963–97

1 FOOD AND KINDRED PRODUCTS
    0.341

2 TEXTILE MILL PRODUCTS
    0.221

3 CHEMICALS AND ALLIED PRODUCTS

    4 Chemicals, except drugs and medicines

        5   Basic industrial inorganic and organic chemistry

           6 Industrial inorganic chemistry

           –0.605

           Industrial organic chemistry

           0.839

        8   Plastics materials and synthetic resins

           –0.302

        9   Agricultural chemicals

           0.244

      10   All other chemicals

           –0.094

           11 Soaps, detergents, cleaners, perfumes, cosmetics and toiletries

           12 Paints, varnishes, lacquers, enamels, and allied products

           13 Miscellaneous chemical products

    14 Drugs and medicines

        –0.429

15   PETROLEUM AND NATURAL GAS EXTRACTION AND REFINING
        0.889

16   RUBBER AND MISCELLANEOUS PLASTICS PRODUCTS
        –0.162

17   STONE, CLAY, GLASS AND CONCRETE PRODUCTS
        0.771

18   PRIMARY METALS

    19 Primary ferrous products
        0.210

    20 Primary and secondary non-ferrous metals
        0.742

21   FABRICATED METAL PRODUCTS
        –0.502

22   MACHINERY, EXCEPT ELECTRICAL

23   Engines and turbines
>   0.373

24   Farm and garden machinery and equipment
>   0.676

25   Construction, mining and material-handling machinery and equipment
>   0.108

26   Metal-working machinery and equipment
>   0.180

27   Office computing and accounting machines
>   0.176

28   Other machinery, except electrical

>   29 Special industry machinery, except metal-working
>   0.320

>   30 General industrial machinery and equipment
>   0.412

>   31 Refrigeration and service industry machinery
>   0.026

>   32 Miscellaneous machinery, except electrical
>   0.525

33   ELECTRICAL AND ELECTRONIC MACHINERY, EQUIPMENT AND SUPPLIES

34   Electrical equipment, except communications equipment

>   35   Electrical transmission and distribution equipment
>   0.100

>   36   Electrical industrial apparatus
>   −0.699

>   37   Other electrical machinery, equipment and supplies
>   0.028

>   *38 Household appliances*

>   *39 Electrical lighting and wiring equipment*

>   *40 Miscellaneous electrical machinery, equipment and supplies*

41   Communications equipment end electronic components

>   42   Radio and television-receiving equipment
>   0.837

>   43   Electronic components and accessories and communications equipment
>   0.204

228

Index

Note: the notes at the ends of chapters have not been indexed.